Monitoring of Cerebral Blood Flow and Metabolism in Intensive Care

Edited by

A. W. Unterberg, G.-H. Schneider, W. R. Lanksch

Acta Neurochirurgica
Supplementum 59

Springer-Verlag Wien New York

Professor Dr. Andreas W. Unterberg
Dr. Gerd-Helge Schneider
Professor Dr. Wolfgang R. Lanksch
Abteilung für Neurochirurgie
Universitätsklinikum Rudolf Virchow
Freie Universität Berlin, Federal Republic of Germany

Typesetting: Best-set Typesetter Ltd., Hong Kong

Printed on acid-free and chlorine free bleached paper

With 76 Figures

Monitoring of cerebral blood flow and metabolism in intensive care:
  proceedings of an international symposium, Berlin, October 1992 / edited by A.W. Unterberg, G.H. Schneider, W.R. Lanksch.
    p. cm. – (Acta neurochirurgica. Supplementum ; 59)
  ISBN-13: 978-3-7091-9304-4
  1. Cerebral ischemia – Diagnosis – Congresses. 2. Cerebral circulation – Measurement – Congresses. 3. Brain – Metabolism – Measurement – Congresses. 4. Patient monitoring – Congresses. 5. Neurological intensive care – Congresses. I. Unterberg, A. (Andreas) II. Schneider, G. H. (Gerd Helge), 1945– . III. Lanksch, Wolfgang, 1938– . IV. Series.
  [DNLM: 1. Cerebrovascular Circulation – physiology – congresses. 2. Brain – blood supply – congresses. 3. Brain – metabolism – congresses. 4. Monitoring, Physiologic – congresses. 5. Cerebrovascular Disorders – physiopathology – congresses. 6. Intensive Care – congresses. W1 AC8661 no. 59 1993 / WL 302 M744 1993]
  IN PROCESS 616.8′1—dc20 DNLM/DLC 93-23371 CIP for Library of Congress.

ISSN 0065-1419
ISBN-13: 978-3-7091-9304-4   e-ISBN-13: 978-3-7091-9302-0
DOI: 10.1007/978-3-7091-9302-0

Preface

Until recently, monitoring of cerebral blood flow and metabolism was an unattained goal. Determination of cerebral blood flow was limited to intermittent measurements and particularly difficult to perform in critically ill patients. Meanwhile there are techniques available, however, to monitor cerebral blood flow and cerebral oxygenation, both globally and regionally.

Therefore we thought it worthwhile to discuss these new continuous techniques and to compare them with well-known techniques which discontinuously measure CBF. For that purpose, an international workshop with some leading experts in the field was held in October 1992 in Berlin. The workshop consisted of about 20 lectures, either reviews on a special topic, or latest results. These contributions were given by invitation and were extensively discussed. Unfortunately it is impossible to reproduce the discussions. On the other hand, all speakers delivered a manuscript promptly after the meeting so that we were able to edit them within a short time. Since monitoring of cerebral blood flow in intensive care is a rapidly growing and changing topic, the written contributions should be quickly available. Authors, editors and publishers have tried to come close to this ideal. As editors we would like to thank the authors and the publishers who enabled us to come out with this volume of the proceedings as early as possible.

This volume is arranged according to the workshop, basic physiological and pathophysiological reviews in the beginning, followed by reports on the role of intermittent measurements of cerebral blood flow and metabolism in critically ill patients. Thereafter, various techniques of continuous monitoring of cerebral blood flow and metabolism are presented, such as CBF-monitoring by thermodiffusion, tissue pO_2-measurements, laser doppler spectroscopy, near infrared spectroscopy, dopplersonography and jugular venous oximetry. Since monitoring of oxygen saturation in the jugular bulb to measure global cerebral oxygenation has been taken up by several groups recently, this topic is certainly the most intensely discussed.

Though it is extremely difficult to summarize a meeting, there are some remarkable results which shall briefly be mentioned. With the introduction of ICP-monitoring in neurosurgical intensive care, a main aim of treatment was to control ICP below 20 mmHg. Later, the necessity of a sufficient cerebral perfusion pressure (of at least 50 mmHg) was stressed. Today, it becomes more and more evident (by various monitoring techniques) that following brain injury a higher cerebral perfusion pressure is necessary to secure adequate perfusion and oxygenation. The new CPP limit is probably in the range of 70 mmHg, but there are marked individual differences. Another point which was stressed by several authors is the potentially harmful effect of hyperventilation on cerebral blood flow and oxygenation. Great caution should be taken when hyperventilation is used to control increased intracranial pressure. Therefore, most authors advocate not to hyperventilate below a pCO_2 of 30 mmHg, unless cerebral oxygenation is monitored.

As to the various methods of monitoring which were discussed controversially, it is clear that we are still lacking the optimal monitoring technique which yields continuous data on CBF and oxygenation in various brain regions. There has definitely been progress, particularly by jugular venous oximetry. Moreover, less invasive techniques such as near infrared spectroscopy, appear to be very promising for the near future. The optimal monitoring, e.g. continuous PET, will remain utopian. Nevertheless, we are expecting considerable progress in the next few years, which hopefully will contribute to a better outcome for critically ill patients.

Finally, we should like to express our gratitude to the Deutsche Forschungsgemeinschaft, Bonn and to the Senatsverwaltung für Wissenschaft und Forschung, Berlin, for generous funding of the meeting.

A. W. Unterberg
G.-H. Schneider
W. R. Lanksch

Contents

Physiology and Pathophysiology of
Cerebral Blood Flow and Metabolism

Acta Neurochir (1993) [Suppl] 59: 3–10

Regulation of Cerebral Blood Flow – A Brief Review

M. Wahl and **L. Schilling**

Department of Physiology, University of Munich, Federal Republic of Germany

Summary

Cerebral blood flow is largely independent of perfusion pressure when autoregulation is intact. Cerebral circulation is regulated mainly by changes of vascular resistance. Resistance can be modulated by local-chemical and endothelial factors, by autacoids, and by release of transmitters from perivascular nerves.

Local-chemical factors such as H^+-, K^+-, Ca^{2+}-ions, adenosine, and osmolarity are involved in the regulation of cerebrovascular resistance during cortical activation and under pathological conditions such as hypoxia or ischaemia.

Endothelial factors such as thromboxane A_2, endothelin (ET), endothelium derived constrictor factor and endothelium derived relaxing (EDRF, identified as nitric oxide, NO) or hyperpolarizing (EDHF) factor, and prostacyclin (PGI_2), can be released by physical stimuli such as shear stress or haemorrhage, by autacoids, by neurotransmitters, and by cytokines. Several of these factors (NO, PGI_2, ET) can also be released from neurons and astrocytes thus enabling a coupling between parenchymal function and flow.

Autacoids like histamine, bradykinin, eicosanoids, and free radicals influence cerebrovascular resistance, capacitance vessels and the permeability of the blood-brain barrier under pathological conditions. They are released by trauma, ischaemia, seizures and inflammation.

Cerebral arteries are innervated by several systems. The sympathetic-noradrenergic fibres originate from the superior cervical ganglion. By releasing the constricting transmitters norepinephrine and neuropeptide Y this system extends the range of autoregulation. The parasympathetic cholinergic system with the dilating transmitters acetylcholine and vasoactive intestinal polypeptide may prevent ischaemia. Besides the intracerebral noradrenergic and serotonergic perivascular innervation with an unclear function, a trigeminal innervation has been described. Its dilating transmitters substance P, calcitonin gene-related peptide, and neurokinin A can be released in an axon reflex-like manner. This system may be involved in vascular headache or vasospasm.

In conclusion, the regulation of cerebral circulation involves the same mechanisms as in peripheral organs.

Keywords: Local-chemical factors; endothelical factors; autacoids; vascular nerves.

Introduction

According to Ohm's law blood flow is dependent on perfusion pressure and vascular resistance. Cerebral blood flow (CBF), however, is independent of the cerebral perfusion pressure because of autoregulation in a range between 80 and 180 mmHg[41, 78, 80]. Myogenic mechanisms are involved in the autoregulatory response with constriction and dilatation during increase and decrease of perfusion pressure, respectively. This could be demonstrated in isolated cerebral arteries in which local-chemical, humoral and neural mechanisms were excluded[27]. It is discussed controversially whether the presence of endothelium is a prerequisite for autoregulation[78]. In situ, local-chemical factors like H^+, K^+ and adenosine are unimportant for autoregulation[80]. On the other hand, changes in sympathetic tone extend the limits of autoregulation[13, 41, 78, 80].

Under physiological conditions, CBF is mainly regulated by changes in the resistance of cerebral arteries. Resistance can be changed by local-chemical and endothelial factors, by autacoids, and by perivascular nerves.

Local-Chemical (Parenchymal) Factors

Local-chemical regulation of vascular resistance means that depending on the function and the metabolism of the parenchymal cells several compounds are released or taken up, thus inducing a change in the composition of the interstitial fluid. This may induce a change of the membrane potential and/or a change in the membrane conductance leading to relaxation or contraction of smooth muscle cells. Such a mechanism would enable a regionally tight coupling between parenchymal cell function and blood supply.

Employing perivascular microapplication the vasomotor effects of several local-chemical factors

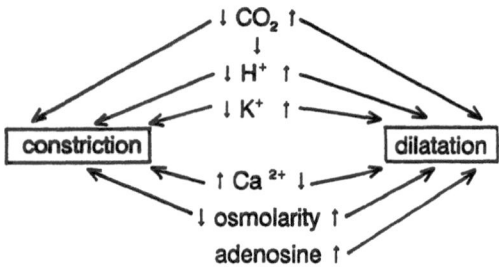

Fig. 1. Local-chemical factors

Table 1. *Interactions of Local Chemical Factors*

1. K^+ dominates over H^+ and adenosine
2. H^+ dominates over adenosine
3. $Ca^{++} \downarrow$ increases dilatation induced by $K^+ \uparrow$ or $H^+ \uparrow$
4. $Ca^{++} \uparrow$ increases constriction incuded by $H^+ \downarrow$

Table 2. *Participation of Local Factors in the Control of CBF under Several Conditions*

1. Cortical activation
   A) slight: $K^+ \uparrow$, $Ca^{++} \downarrow$
   B) strong: $K^+ \uparrow$, $Ca^{++} \downarrow$, $H^+ \uparrow$, adenosine $\uparrow$

2. Autoregulation
   No change of H^+, K^+, adenosine

3. Strong hypoxia or ischaemia
   $K^+ \uparrow$, $H^+ \uparrow$, adenosine $\uparrow$

have been demonstrated as shown in Fig. 1[41, 80]. During simultaneous concentration changes of several local factors their interactions were studied which are listed in Table 1[41, 80]. These results demonstrate that H^+-ions, K^+-ions, Ca^{2+}-ions, adenosine, and osmolarity of the extracellular fluid are vasoactive factors acting on cerebral arteries. The quantitative role of these factors for the regulation of vascular resistance depends on their interactions. Employing microelectrodes, microdialysis technique and tissue analysis the concentrations of H^+, K^+, Ca^{2+} and adenosine in the interstitial space and in brain tissue were found to change during several experimental conditions as shown in Table 2[78, 80, 82]. In summary, it can be concluded that several local-chemical factors are involved in the regulation of cerebrovascular resistance during cortical activation and under pathological conditions such as hypoxia or transient ischaemia. Under certain circumstances, such as cortical spreading depression, however, their vasomotor effects may be at-

tenuated[83]. In contrast, local factors do not appear to be important for cerebral autoregulation.

Endothelial Factors

The concept that the endothelium is important for mediation of vasomotor responses elicited by several agents was introduced by Furchgott and Zawadzki[19]. They found that the dilating effect of acetylcholine (ACh) is endothelium dependent. Employing the sandwich-technique the release of the endothelium derived relaxing factor (EDRF) which is nitric oxide (NO)[28] or a NO-containing moiety[46] was detected. The mechanisms of NO release from the endothelial cell[50] and its action in the smooth muscle cell are shown schematically in Fig. 2. Formation of c GMP relaxes the smooth muscle cell by decreasing the intracellular Ca^{2+} concentration and/or the Ca^{2+} sensitivity of the contractile proteins[33]. Meanwhile it has been found that several constrictor and dilating factors can be released from the endothelium as schematically shown in Fig. 3.

Constrictor factors are endothelin (ET), thromboxane $A_2 (TxA_2)$, and the hypothetical endothelium derived constrictor factor (EDCF). Dilating factors are EDRF, prostacyclin (PGI_2), and a postulated endothelium derived hyperpolarizing factor

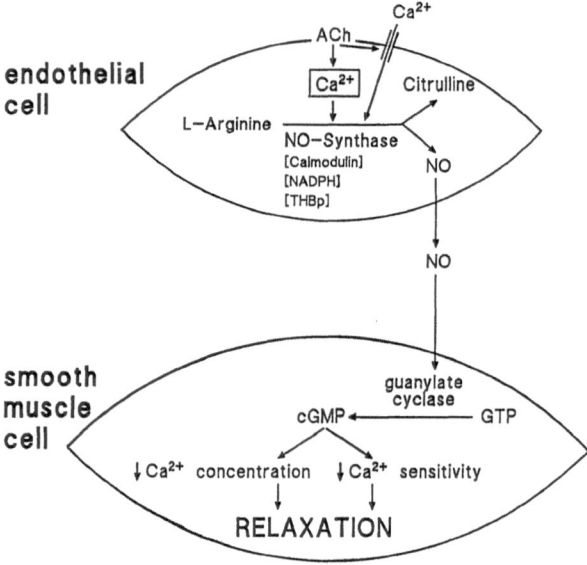

Fig. 2. Activation of endothelial nitric oxide synthase (*NO-Synthase*) by acetylcholine (*ACh*) and of smooth muscle soluble guanylate cyclase. *NO*: Nitric oxide; *NADPH*: reduced nicotine-amide-adenine-nucleotide phosphate; *cGMP*: cyclic guanosine monophosphate; *THBp*: tetrahydrobiopterin; *GTP*: guanosine triphosphate

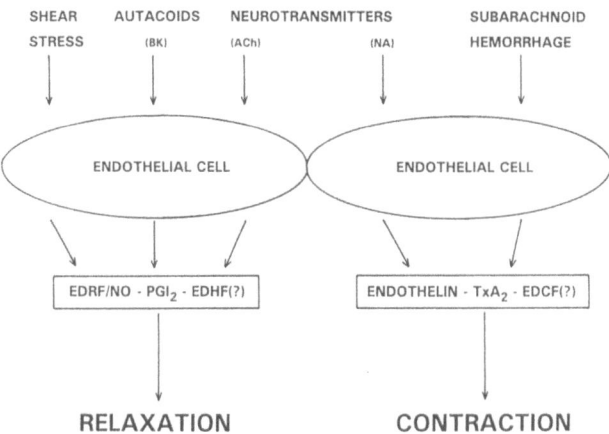

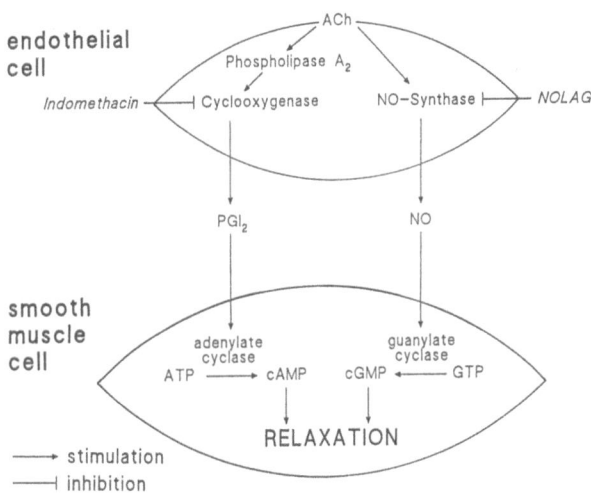

Fig. 3. Factors inducing smooth muscle cell relaxation or contraction by release of endothelial relaxing or constrictor factors. *BK*: Bradykinin; *ACh*: acetylcholine; *NA*: noradrenaline; *EDRF/NO*: endothelium-derived relaxing factor/nitric oxide; *EDHF*: endothelium-derived hyperpolarizing factor; *TxA$_2$*: thromboxane A$_2$; *EDCF*: endothelium-derived constrictor factor; *PGI$_2$*: prostacyclin

Fig. 4. Mechanisms mediating the relaxant effect of acetylcholine (*ACh*). *NO-synthase*: nitric oxide synthase; *NOLAG*: N^G-nitro-L-arginine; *PGI$_2$*: prostacyclin; *NO*: nitric oxide; *ATP*: adenosine triphosphate; *cGMP*: cyclic guanosine monophosphate; *GTP*: guanosine triphosphate; *cAMP*: cyclic adenosine monophosphate

(EDHF). These endothelial factors can be released by different stimuli such as shear stress, autacoids and neurotransmitters. For instance, EDRF can be released by neurotransmitters such as ACh and substance P, or autacoids such as bradykinin or 5-hydroxytryptamine (5-HT) or by physical stimuli such as shear stress.

A novel potent vasoconstrictor peptide, ET, increases via ET$_A$ receptors on smooth muscle cells Ca^{2+} influx and Ca^{2+} release from intracellular stores. Under in vitro[25, 31, 36] and in situ[14, 59] conditions ET has been found to be a potent constrictor of pial, intraparenchymal and basilar arteries. Because of its long lasting vasomotor response it is discussed as one causal factor of cerebral vasospasm following subarachnoid haemorrhage[47]. Besides the hypothetical EDCF[35] an endothelial production and release of TxA$_2$ has been found in isolated canine cerebral arteries. TxA$_2$ obviously mediates the constriction induced by noradrenaline and ATP[67, 68, 79].

As shown in Fig. 4 ACh is shown to release EDRF and PGI$_2$ from endothelial cells. EDRF (NO) mediates the ACh induced relaxation of middle cerebral and basilar arteries under in vitro[34, 55, 62] and in situ[15, 48] conditions. Similarly, the dilating response of the smaller pial arteries to ACh[38, 46] and 5-HT[56] is induced by this pathway. Whereas the dilating response of the basilar arteries of dogs and rats to bradykinin is mediated by NO[34, 48] a different

endothelial factor is involved in the dilating effect of bradykinin in pial arteries of other species[38, 46, 56]. Although the resting tone of the middle cerebral and basilar artery is, at least in part, influenced by a continuous release of NO under in situ[15, 16] and in vitro[55, 62] conditions, this could not be confirmed in pial arteries in situ[8, 16, 22, 56]. Furthermore, a decrease of CBF has been found in most studies after application of an inhibitor of NO-synthase[17, 39, 40, 53, 73, 88].

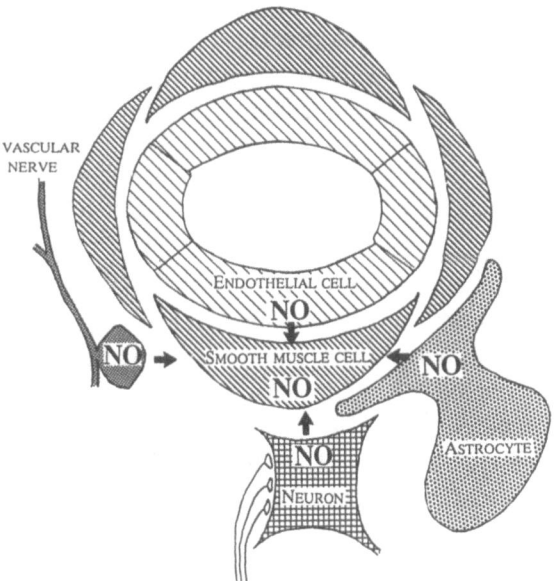

Fig. 5. Sources of nitric oxide (*NO*) in the brain

As shown in Fig. 5 NO is not only released from the endothelium but can also be formed in the smooth muscle cell under certain conditions[9, 64]. Since NO or a compound similar to EDRF is also released from astrocytes and neurons[6, 7, 18, 20, 21, 52, 65] and from perivascular nerves[42, 75] it may act like a local-chemical factor and a neurotransmitter, respectively. Employing the model of cortical spreading depression (CSD) we have tested the hypothesis whether NO acts like a local-chemical factor and mediates coupling between neuronal activity and vascular resistance. CSD is characterized by a propagating wave of transient neuronal activation indicated by a negative DC-shift and accompanied by a transient pial arterial dilatation[83]. By infusing 10^{-4}M N^G-Nitro-L-Arginine, a NO-synthase inhibitor, over a period of up to 4.5 min into the perivascular space of individual arteries the transient dilatation of these arteries during CSD was reduced by about 50%[83a]. This is in accordance with recent findings in other models of cortical activation[1, 53]. It demonstrates that NO is an additional factor besides classical local-chemical factors, trigeminal nerves, and perhaps prostanoids which appear to be involved in the mediation of functional hyperaemia[82].

PGI_2 is another dilating compound which can be formed and released by endothelial cells[30]. PGI_2 and NO are both involved in the mediation of ACh induced dilatation of isolated rabbit basilar artery[62]. This conclusion is based on the finding that the response to ACh could only partially be blocked by indomethacin and N^G-Nitro-L-Arginin, inhibitors of cyclo-oxygenase and NO-synthase, respectively. However, the relaxation was completely suppressed in the presence of both inhibitors. Species and vessel differences appear to be important since the dilating effect of ACh in isolated rat basilar and rabbit middle cerebral arteries is exclusively mediated by NO[45, 55].

EDHF may act upon ATP-sensitive K^+ channels in the smooth muscle cell membrane inducing hyperpolarization and closure of voltage dependent Ca^{2+} channels[74]. It appears to participate in the mediation of the ACh induced dilatation of isolated rabbit middle cerebral arteries[4, 5]. The functional meaning of this mechanism has been questioned by others[55, 62]. Furthermore, an uncoupling between hyperpolarization and dilatation due to ACh has been found in the rabbit basilar artery also questioning a causal role of EDHF[58].

Autacoids

Several autacoids appear to influence cerebrovascular resistance, capacitance vessels and permeability of the blood-brain barrier under several pathological conditions (Tables 3 and 4)[63, 78, 80, 84, 86, 87]. Histamine can be released from mast cells or from the vessel wall itself. Furthermore, it can act as a neurotransmitter or transmitter of perivascular nerves[63]. During perivascular microapplication histamine dilates extraparenchymal arteries via activation of H_2 receptors. Intravascularly applied histamine does not changes CBF since it does not penetrate the blood-brain barrier. However, intravascular and extravascular application of histamine can increase blood-brain barrier permeability via activation of H_2-receptors and induce brain oedema[63, 80, 86].

Bradykinin can also be released from brain parenchyma since all components of an intracerebral kallikrein kinin system have been found[84, 87]. Opening of the blood-brain barrier following a cryogenic lesion induces an uptake of blood born kininogen and a release of bradykinin. Bradykinin dose-dependently dilates cerebral arteries via B_2-kininergic receptors when acting from the perivascular side. The vasomotor effect of bradykinin is endothelium-dependent and can be mediated by various endothelial factors[84]. The B_2-receptor which induces dilatation and increase of CBF can only be reached from the parenchymal side. In contrast, opening of the blood-brain barrier for sodium fluorescein via endothelial B_2-receptors can be induced by intraarterial infusion and cortical superfusion of bradykinin. As cold lesion induced brain swelling can be reduced by infusion of the kallikrein inhibitor aprotinin bradykinin appears to be one

Table 3. *Autacoids*

Bradykinin	Leukotrienes
Histamine	Free radicals
Arachidonic acid	Serotonin
Prostanoids	

Table 4. *Release of Autacoids*

Brain injury
Ischaemia
Seizure
Inflammation

mediator of brain oedema and secondary brain damage after brain injury[78, 84, 86, 87].

Eicosanoids and their precursor arachidonic acid are released in brain parenchyma and in part by the vessel wall itself under pathological conditions (Table 4)[86]. Cortical superfusion with arachidonic acid induces slight vasomotor responses only but it opens the blood-brain barrier for small and large tracers and induces oedema[78, 86]. Therefore, arachidonic acid appears to be an important mediator of vasogenic brain oedema besides bradykinin[84, 86, 87], histamine[63, 86] and free radicals[86]. In contrast, leukotrienes which are formed from arachidonic acid by the lipoxygenase pathway do not increase the permeability of the blood-brain barrier in intact vessels[77]. Leukotrienes as well as TxA_2 are potent constrictors of cerebral arteries[77] and appear to be involved in the development of cerebral vasospasm after subarachnoid haemorrhage. Surprisingly, TxA_2 and the stable synthetic analogue U46619 are less potent in situ compared to in vitro conditions[85, 89]. The balance between dilating (PGE_2 and PGI_2) and constrictor prostanoids is of importance for the regulation of vascular resistance under pathological conditions. It has been discussed that the development of vasospasm after subarachnoid haemorrhage is at least in part mediated by the reduced release of PGI_2 and by an increased formation of leukotrienes and TxA_2[78, 81].

Perivascular Nerves

Cerebral arteries are innervated by several systems which are listed in Fig. 6. The sympathetic noradrenergic fibres originating in the superior cervical ganglion release the constrictor transmitters norepinephrine (NE) and neuropeptide Y (NPY)[12, 13, 29, 41, 44, 76, 80]. Activation of sympathetic fibres reduces CBF which has most convincingly been shown in awake animals. During sympathetic stimulation the upper limit of autoregulation is shifted to a higher blood pressure and opening of the blood-brain barrier is prevented. Conversely, a reduction of the sympathetic discharge extends the lower limit of autoregulation to lower values of blood pressure. Furthermore, sympathetic fibres seem to participate in the development of vasospasm after subarachnoid haemorrhage[44, 78, 81].

The parasympathetic cholinergic fibres originating mainly in the sphenopalatine and otic ganglia[12,

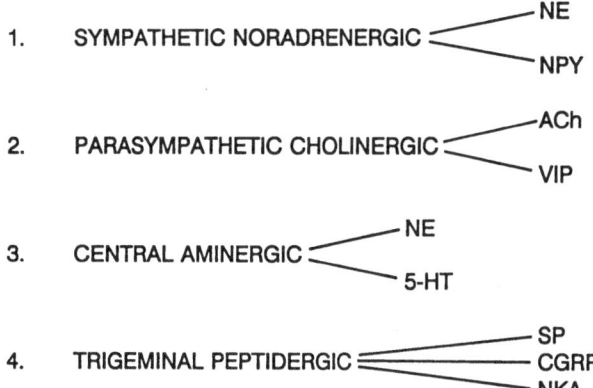

Fig. 6. Innervation of cerebral vessels. *NE*: Norepinephrine; *NPY*: neuropeptide Y; *ACh*: acetylcholine; *VIP*: vasoactive intestinal polypeptide; *5-HT*: 5-hydroxytryptamine; *SP*: substance P; *CGRP*: calcitonin gene related peptide; *NKA*: neurokinin A

23, 24, 69, 72] increase CBF[54, 66, 70] by release of the dilating transmitters ACh and vasoactive intestinal polypeptide (VIP) which are mostly colocalized[12, 23, 54]. A coupling between neuronal function and blood flow has been suggested for this system[43]. Recent findings demonstrate that parasympathetic nerves may reduce infarct size after occlusion of the middle cerebral artery[32, 37].

Several aminergic fibres with the transmitters NE or 5-HT appear to connect the brain stem with neurons and blood vessels of the cortex[3, 44, 57]. On the other hand, the serotonergic innervation of cerebral arteries has been questioned by others[10, 60]. Although these systems have been described as capable of changing vascular permeability[57] or eliciting vasomotor responses[2, 3], respectively, their functional significance is still unclear.

Extraparenchymal cerebral arteries are also innervated by fibres mainly originating from the trigeminal ganglion and releasing the dilating transmitters substance P (SP), calcitonin gene related peptide (CGRP) and neurokinin A (NKA)[24, 26, 49, 71]. This system contributes to the increase of CBF during cortical activation[61] and is discussed as being involved in the mediation of vascular headache[51]. Furthermore, a depletion of the dilating transmitters after subarachnoid haemorrhage appears to participate in the development of vasospasm[11].

In conclusion, in the regulation of the cerebral microcirculation the same principles and mechanisms are involved as in other organs, i.e. cardiac or skeletal muscles. The quantitative role of the individual factors for the regulation of regional cerebral blood flow, however, is different from other organs.

References

1. Adachi T, Inanami O, Saito A (1992) Nitric oxide (NO) is involved in increased cerebral cortical blood flow following stimulation of the nucleus basalis of Meynert in anesthetized rats. Neurosci Lett 139: 201–204
2. Bonvento G, Lacombe P, Seylaz J (1989) Effects of electrical stimulation of the dorsal raphe nucleus on local cerebral blood flow in the rat. J Cereb Blood Flow Metab 9: 251–255
3. Bonvento G, MacKenzie ET, Edvinsson L (1991) Serotonergic innervation of the cerebral vasculature, relevance to migraine and ischemia. Brain Res Rev 16: 257–263
4. Brayden JE (1990) Membrane hyperpolarization is a mechanism of endothelium-dependent cerebral vasodilation. Am J Physiol 259: H668–H673
5. Brayden JE, Wellman GC (1989) Endothelium-dependent dilation of feline cerebral arteries: role of membrane potential and cyclic nucleotides. J Cereb Blood Flow Metab 9: 256–263
6. Bredt DS, Hwang PM, Snyder SH (1990) Localisation of nitric oxide synthase indicating a neural role for nitric oxide. Nature 347: 768–769
7. Bredt DS, Snyder SH (1989) Nitric oxide mediates glutamate-linked enhancement of cGMP levels in the cerebellum. Proc Natl Acad Sci USA 86: 9030–9033
8. Busija DW, Leffler CW, Wagerle LC (1990) Mono-L-arginine containing compounds dilate piglet pial arterioles via an endothelium derived relaxing factor like substance. Circ Res 67: 1374–1380
9. Busse R, Mülsch A (1990) Induction of nitric oxide synthase by cytokines in vascular smooth muscle cells. FEBS Lett 275: 87–90
10. Chang J-Y, Hardebo JE, Owman C (1990) Kinetic studies on uptake of serotonin and noradrenaline into pial arteries of rats. J Cereb Blood Flow Metab 10: 22–31
11. Edvinsson L, Delgado-Zygmunt T, Ekman R, Jansen I, Svendgaard N-A, Uddman R (1990) Involvement of perivascular sensory fibers in the pathophysiology of cerebral vasospasm following subarachnoid hemorrhage. J Cereb Blood Flow Metab 10: 602–607
12. Edvinsson L, Hara H, Uddman R (1989) Retrograde tracing of nerve fibers to the rat middle cerebral artery with true blue: colocalization with different peptides. J Cereb Blood Flow Metab 9: 212–218
13. Edvinsson L, MacKenzie ET (1977) Amine mechanisms in the cerebral circulation. Pharmacol Rev 28: 275–348
14. Faraci FM (1989) Effects of endothelin and vasopressin on cerebral blood vessels. Am J Physiol 257: H799–H803
15. Faraci FM (1990) Role of nitric oxide in regulation of basilar artery tone in vivo. Am J Physiol 259: H1216–H1221
16. Faraci FM (1991) Role of endothelium-derived relaxing factor in cerebral circulation: large arteries vs. microcirculation. Am J Physiol 261: H1038–H1042
17. Faraci FM, Heistad DD (1992) Endothelium-derived relaxing factor inhibits constrictor responses of large cerebral arteries to serotonin. J Cereb Blood Flow Metab 12: 500–506
18. Förstermann U, Schmidt HHHW, Pollock JS, Heller M, Murad F (1991) Enzymes synthesizing guanylate cyclase-activating factors in endothelial cells, neuroblastoma cells, and rat brain. J Cardiovasc Pharmacol 17 [Suppl 3]: S57–S64
19. Furchgott RF, Zawadzki JV (1980) The obligatory role of endothelial cells in the relaxation of arterial smooth muscle by acetylcholine. Nature 288: 373–376
20. Garthwaite J (1991) Glutamate, nitric oxide and cell-cell signalling in the nervous system. Trends Neurosci 14: 60–67
21. Garthwaite J, Garthwaite G, Palmer RMJ, Moncada S (1989) MNDA receptor activation induces nitric oxide synthesis from arginine in rat brain slices. Eur J Pharmacol 172: 413–416
22. Haberl RL, Decker PJ, Piepgras A, Einhäupl K (1991) Is L-arginine the precursor of an endothelium-derived relaxing factor in the cerebral microcirculation? J Cardiovasc Pharmacol 17 [Suppl 3]: S15–S18
23. Hara H, Jansen I, Ekman R, Hamel E, MacKenzie ET, Uddman R, Edvinsson L (1989) Acetylcholine and vasoactive intestinal peptide in cerebral blood vessels: effect of extirpation of the sphenopalatine ganglion. J Cereb Blood Flow Metab 9: 204–211
24. Hardebo JE, Arbab M, Suzuki N, Svendgaard NA (1991) Pathways of parasympathetic and sensory cerebrovascular nerves in monkeys. Stroke 22: 331–342
25. Hardebo JE, Kahrström J, Owman C, Salford LG (1989) Endothelin is a potent constrictor of human intracranial arteries and veins. Blood Vessels 26: 249–253
26. Hardebo JE, Suzuki N, Owman C (1989) Origins of substance P- and calcitonin gene – related peptide-containing nerves in the internal carotid artery of rat. Neurosci Lett 101: 39–45
27. Harder DR (1984) Pressure-dependent membrane depolarization in cat middle cerebral artery. Circ Res 55: 197–202
28. Ignarro LJ, Buga GM, Wood KS, Byrns RE, Chaudhuri G (1987) Endothelium-derived relaxing factor produced and released from artery and vein is nitric oxide. Proc Natl Acad Sci USA 84: 9265–9269
29. Itakura T, Nakakita K, Kamei I (1987) Immunohistochemical studies of peptidergic nerve fibres in the cerebral vein. In: Edvinsson L, et al (eds) Peptidergic mechanisms in the cerebral circulation. Horwood, Chichester, pp 34–47
30. Jaiswal N, Jaiswal RK, Malik KU (1991) Muscarinic receptor-mediated prostacyclin and cGMP synthesis in cultured vascular cells. Mol Pharmacol 40: 101–106
31. Jansen I, Fallgren B, Edvinsson L (1989) Mechanisms of action of endothelin on isolated feline cerebral arteries: in vitro pharmacology and electrophysiology. J Cereb Blood Flow Metab 9: 743–747
32. Kano M, Moskowitz MA, Yokota M (1991) Parasympathetic denervation of rat pial vessels significantly increases infarction volume following middle cerebral artery occlusion. J Cereb Blood Flow Metab 11: 628–637
33. Karaki H (1989) Ca^{2+} localization and sensitivity in vascular smooth muscle. Trends Pharmacol Sci 10: 320–324
34. Katusic ZS, Marshall JJ, Kontos HA, Vanhoutte PM (1989) Similar responsiveness of smooth muscle of the canine basilar artery to EDRF and nitric oxide. Am J Physiol 257: H1235–H1239
35. Katusic ZS, Shepherd JT, Vanhoutte PM (1988) Endothelium-dependent contractions to calcium ionophore A23187, arachidonic acid, and acetylcholine in canine basilar arteries. Stroke 19: 476–479
36. Kauser K, Rubanyi GM, Harder DR (1990) Endothelium-dependent modulation of endothelin-induced vasoconstriction and membrane depolarization in cat cerebral arteries. J Pharmacol Exp Ther 252: 93–97

37. Koketsu N, Moskowitz MA, Kontos HA, Yokata M, Shimizu T (1992) Chronic parasympathetic sectioning decreases regional cerebral blood flow during hemorrhagic hypotension and increases infarct size after middle cerebral artery occlusion in spontaneously hypertensive rats. J Cereb Blood Flow Metab 12: 613–620

38. Kontos HA, Wei EP, Kukreja RC, Ellis EF, Hess ML (1990) Differences in endothelium – dependent cerebral dilation by bradykinin and acetylcholin. Am J Physiol 258: H1261–H1266

39. Kovách AGB, Szabó C, Benyó Z, Csáki C, Greenberg JH, Reivich M (1992) Effects of N^G-nitro-L-arginine and L-arginine on regional cerebral blood flow in the cat. J Physiol 449: 183–196

40. Kozniewska E, Oseka M, Stys T (1992) Effects of endothelium derived nitric oxide on cerebral circulation during normoxia and hypoxia in the rat. J Cereb Blood Flow Metab 12: 311–317

41. Kuschinsky W, Wahl M (1978) Local chemical and neurogenic regulation of cerebral vascular resistance. Physiol Rev 58: 656–689

42. Lee TJ-F, Sarwinski SJ (1991) Nitric oxidergic neurogenic vasodilatation in the porcine basilar artery. Blood Vessels 28: 407–412

43. Lou HC, Edvinsson L, MacKenzie ET (1987) The concept of coupling blood flow to brain function: revision required? Ann Neurol 22: 289–297

44. MacKenzie ET, Scatton B (1987) Cerebral circulatory and metabolic effects of perivascular neurotransmitters. Crit Rev Clin Neurobiol 2: 357–419

45. Mackert JRL, Parsons AA, Ksoll E, Schilling L, Wahl M (1990) Methylene blue and N^G-nitro-L-arginine inhibition of acetylcholine induced relaxation of rat isolated basilar artery. Int J Microcirc Clin Exp 9: 227

46. Marshall J, Kontos HA (1990) Endothelium-derived relaxing factors. A perspective from in vivo data. Hypertension 16: 371–386

47. Matsumura Y, Ikegawa R, Suzuki Y, Takaoka M, Uchida T, Kido H, Shinyama H, Hayashi K, Watanabe M, Morimoto S (1991) Phosphoramidon prevents cerebral vasospasm following subarachnoid hemorrhage in dogs: The relationship to endothelin-1 levels in the cerebrospinal fluid. Life Sci 49: 841–848

48. Mayhan WG (1990) Impairment of endothelium-dependent dilatation of basilar artery during chronic hypertension. Am J Physiol 259: H1455–H1462

49. McCulloch J, Edvinsson L (1987) Calcitonin gene-related peptide and the trigeminal innervation of the cerebral vasculature. In: Edvinsson L, et al (eds) Peptidergic mechanisms in the cerebral circulation. Horwood, Chichester, pp 132–151

50. Moncada S, Palmer RMJ, Higgs EA (1991) Nitric oxide: physiology, pathophysiology and pharmacology. Pharmacol Rev 43: 109–142

51. Moskowitz MA (1989) Trigeminovascular system: form and function in the cephalic vasculature. In: Seylaz J, et al (eds) Neurotransmission and cerebrovascular function II. Elsevier, Amsterdam, pp 311–328

52. Murphy S, Minor RL Jr, Welk G, Harrison DG (1990) Evidence for an astrocyte-derived vasorelaxing factor with properties similar to nitric oxide. J Neurochem 55: 349–351

53. Northington FJ, Matherne GP, Berne RM (1992) Competitive inhibition of nitric oxide synthase prevents the cortical hyperemia associated with peripheral nerve stimulation. Proc Natl Acad Sci USA 89: 6649–6652

54. Owman C, Hanko J, Hardebo JE, Kährström J (1986) Neuropeptides and classical autonomic transmitters in the cardiovascular system: existence, coexistence, action, interaction. In: Owman C, et al (eds) Neural regulation of brain circulation. Elsevier, New York, pp 299–331

55. Parsons AA, Schilling L, Wahl M (1991) Analysis of acetylcholine-induced relaxation of rabbit isolated middle cerebral artery: effects of inhibitors of nitric oxide synthesis, Na, K-ATPase, and ATP-sensitive K-channels. J Cereb Blood Flow Metab 11: 700–704

56. Parsons AA, Wang Q, Schilling L, Lassen NA, Wahl M (1991) Effects of N^G-nitro-L-arginine (NOLAG) on rat pial arterioles in situ. Pflügers Arch Eur J Physiol 419 [Suppl 1]: R112

57. Raichle ME, Hartmann BK, Eichling JO, Sharpe LG (1975) Central noradrenergic regulation of cerebral blood flow and vascular permeability. Proc Natl Acad Sci USA 72: 3726–3730

58. Rand VE, Garland CJ (1992) Endothelium-dependent relaxation to acetylcholine in the rabbit basilar artery: importance of membrane hyperpolarization. Br J Pharmacol 106: 143–150

59. Robinson MJ, McCulloch J (1990) Contractile responses to endothelin in feline cortical vessels in situ. J Cereb Blood Flow Metab 10: 285–289

60. Saito A, Lee TJ-F (1987) Serotonin as an alternative transmitter in sympathetic nerves of large cerebral arteries of the rabbit. Circ Res 60: 220–228

61. Sakas DE, Moskowitz MA, Wei EP, Kontos HA, Kano M, Ogilvy CS (1989) Trigeminovascular fibers increase blood flow in cortical gray matter by axon reflex-like mechanisms during acute severe hypertension or seizures. Proc Natl Acad Sci USA 86: 1401–1405

62. Schilling L, Parsons AA, Mackert JRL, Wahl M (1991) Is K^+ channel activation, EDRF, or cyclooxygenase products involved in acetylcholine-induced relaxation of rabbit isolated basilar artery? J Cereb Blood Flow Metab 11 [Suppl 2]: S256

63. Schilling L, Wahl M (1993) Histaminergic effects on cerebral hemodynamics. In: Phillis JW (ed) The regulation of cerebral blood flow. CRC Press, Boca Raton, pp 113–128

64. Schini VB, Vanhoutte PM (1991) L-arginine evokes both endothelium-dependent and endothelium-independent relaxations in L-arginine depleted aortas of the rat. Circ Res 68: 209–216

65. Schmidt HHHW, Wilke P, Evers B, Böhme E (1989) Enzymatic formation of nitrogen oxides from L-arginine in bovine brain cytosol. Biochem Biophys Res Commun 165: 284–291

66. Seylaz J, Hara H, Pinard E, Mraovitch S, MacKenzie ET, Edvinsson L (1988) Effect of stimulation of the sphenopalatine ganglion on cortical blood flow in the rat. J Cereb Blood Flow Metab 8: 875–878

67. Shirahase H, Fujiwara M, Usui H, Kurahashi K (1987) A possible role of thromboxane A_2 in endothelium in maintaining resting tone and producing contractile response to acetylcholine and arachidonic acid in canine cerebral arteries. Blood Vessels 24: 117–119

68. Shirahase H, Usui H, Manabe K, Kurahashi K, Fujiwara M (1988) Endothelium-dependent contraction and endothelium-independent relaxation induced by adenine nucleotides and nucleoside in the canine basilar artery. J Pharmacol Exp Ther 247: 1152–1157

69. Suzuki N, Hardebo J, Owman C (1988) Origins and pathways of cerebrovascular vasoactive intestinal polypeptide-positive nerves in rat. J Cereb Blood Flow Metab 8: 697–712

70. Suzuki N, Hardebo JE, Kahrström J, Owman C (1990) Selective electrical stimulation of postganglionic cerebrovascular parasympathetic nerve fibers originating from the sphenopalatine ganglion enhances cortical blood flow in the rat. J Cereb Blood Flow Metab 10: 383–391

71. Suzuki N, Hardebo JE, Owman C (1989) Trigeminal fibre collarterales storing substance P and calcitonin gene-related peptide associate with ganglion cells containing choline acetyl-transferase and vasoactive intestinal polypeptide in the sphenopalatine ganglion of the rat. An axon reflex modulating parasympathetic ganglion activity? Neuroscience 30: 595–604

72. Suzuki N, Hardebo JE, Owman C (1990) Origins and pathways of choline acetyltransferase-positive parasympathetic nerve fibers to cerebral vessels in rat. J Cereb Blood Flow Metab 10: 399–408

73. Tanaka K, Gotoh F, Gomi S, Takashima S, Mihara B, Shirai T, Nogawa S, Nagata E (1991) Inhibition of nitric oxide synthesis induces a significant reduction in local cerebral blood flow in the rat. Neurosci Lett 127: 129–132

74. Taylor SG, Weston AH (1988) Endothelium-derived hyperpolarizing factor: a new endogenous inhibitor from the vascular endothelium. Trends Pharmacol Sci 9: 272–273

75. Toda N, Okamura T (1991) Role of nitric oxide in neurally induced cerebroarterial relaxation. J Pharmacol Exp Ther 258: 1027–1034

76. Tuor UI, Kelly PAT, Edvinsson L, McCulloch J (1987) Neuropeptide Y: importance in the cerebral circulation. In: Edvinsson L, et al (eds) Peptidergic mechanisms in the cerebral circulation. Horwood, Chichester, pp 75–99

77. Unterberg A, Schmidt W, Wahl M, Ellis EF, Marmarou A, Baethmann A (1991) Evidence against leukotrienes as mediators of brain edema. J Neurosurg 74: 773–780

78. Unterberg A, Wahl M (1992) Regulation of cerebral blood flow: a brief review. In: Schmiedek P, et al (eds) Stimulated cerebral blood flow. Springer, Berlin Heidelberg New York Tokyo, pp 3–11

79. Usui H, Kurahashi K, Shirahase H, Fukui K, Fujiwara M (1987) Endothelium-dependent vasocontraction in response to noradrenaline in the canine cerebral artery. Jpn J Pharmacol 44: 228–231

80. Wahl M (1985) Local chemical, neural, and humoral regulation of cerebrovascular resistance vessels. J Cardiovasc Pharmacol 7 [Suppl 3]: S36–S46

81. Wahl M (1985) A review of neurotransmitters and hormones implicated in mediating cerebral vasospasm. In: Voth D, et al (eds) Cerebral vascular spasm. De Gruyter, New York, pp 223–230

82. Wahl M (1992) Mechanisms of cerebral vasodilatation during neuronal activation by bicuculline: a review. In: Schmiedek P, et al (eds) Stimulated cerebral blood flow. Springer, Berlin Heidelberg New York Tokyo, pp 50–54

83. Wahl M, Lauritzen M, Schilling L (1987) Change of cerebrovascular reactivity after cortical spreading depression in cats and rats. Brain Res 411: 72–80

83a. Wahl M, Parsons AA, Schilling L, Kaumann AJ (1993) Reduction of pial arterial dilatation during cortical spreading depression by application of an inhibitor of nitric oxide synthesis. Cereb Blood Flow Metabol 13 [Suppl 1]: S177

84. Wahl M, Schilling L (1993) Effects of bradykinin in the cerebral microcirculation. In: Phillis JW (ed) The regulation of cerebral blood flow. CRC Press, Boca Raton, pp 315–328

85. Wahl M, Schilling L, Whalley ET (1989) Cerebrovascular effects of prostanoids. Naunyn-Schmiedeberg's Arch Pharmacol 340: 314–320

86. Wahl M, Unterberg A, Baethmann A, Schilling L (1988) Mediators of blood-brain barrier dysfunction and formation of vasogenic brain oedema. J Cereb Blood Flow Metab 8: 621–634

87. Wahl M, Unterberg A, Whalley ET, Baethmann A, Young AR, Edvinsson L, Wagner FFW (1987) Effects of bradykinin on cerebral haemodynamics and blood brain barrier function. In: Edvinsson L, et al (eds) Peptidergic mechanisms in the cerebral circulation. Horwood, Chichester, pp 166–190

88. Wang Q, Paulson OB, Lassen NA (1992) Is autoregulation of cerebral blood flow in rats influenced by nitro-L-arginine, a blocker of the synthesis of nitric oxide? Acta Physiol Scand 145: 297–298

89. Whalley ET, Schilling L, Wahl M (1989) Cerebrovascular effects of prostanoids: in-vitro studies in feline middle cerebral and basilar artery. Prostaglandins 38: 625–634

Correspondence and Reprints: Dr. M. Wahl, Physiologisches Institut, Universität München, Pettenkoferstrasse 12, D-80336 München, Federal Republic of Germany.

Acta Neurochir (1993) [Suppl] 59: 11–17

Relationship of Cerebral Blood Flow Disturbances with Brain Oedema Formation

R. Murr[1], S. Berger[2], L. Schürer[3], O. Kempski[4], F. Staub[2], and A. Baethmann[2]

[1] Institute of Anaesthesiology, [2] Institute of Surgical Research, [3] Department of Neurosurgery, Klinikum Großhadern, Ludwig-Maximilians-University, München, and [4] Institute of Neurosurgical Pathophysiology, University of Mainz, Mainz, Federal Republic of Germany

Summary

Brain oedema is an important factor which compromises maintenance of the cerebral blood flow. Conversely, primary blood flow disturbances are leading to brain oedema. The mechanisms underlying blood flow impairment by brain oedema are associated with an increased regional tissue pressure in proportion to the degree of water accumulation in the parenchyma. The release of vasoactive mediator compounds might be considered in addition. Primary disturbances of the cerebral blood flow, such as focal or global cerebral ischaemia are leading to an increased cerebral water content. A decrease of the cerebral blood flow to ca. 40% of normal or below has been found to result in the development of brain oedema. This flow threshold is in the neighbourhood of the ischaemic flow level causing irreversible tissue damage. Whereas in focal ischaemia oedema formation is a function of the severity of the flow decrease, it is a pathophysiological hallmark of early postischaemic recirculation in global cerebral ischaemia. Nevertheless, during complete interruption of cerebral blood flow translocation of interstitial fluid into the intracellular compartment occurs as manifestation of ischaemic cell swelling. Cell swelling under these conditions may, however, not necessarily indicate cell damage, but more likely a compensatory response attributable to the uptake of excitotoxic transmitters, such as glutamate, and of K^+-ions which are excessively released at the onset of ischaemia into the extracellular space. Purpose of the swelling process, thus, is clearance of extracellular fluid from this material to re-establish homeostasis.

Brain oedema in focal cerebral ischaemia is initially of cytotoxic nature, whereas disruption of the blood-brain barrier evolves with a delay of few days. On the other hand, opening of the blood-brain barrier in global cerebral ischaemia is characteristic of the postischaemic recirculation phase and, probably a function of severity of the reactive hyperaemia. Treatment of cerebral ischaemia, i.e., reestablishment of a normal blood flow to the brain must consider the specific significance of the various pathophysiological phenomena of ischaemic brain oedema, such as cell swelling and opening of the blood-brain barrier.

Keywords: Vasogenic and cytotoxic brain oedema; cerebral ischaemia; ischaemic flow threshold; cell swelling; blood-brain barrier disruption.

Introduction

Relationships between brain oedema formation and disturbances of cerebral blood flow, of the microcirculation, respectively are as close as they are complex. Formation of brain oedema, which is influenced by the actual level of blood flow once the blood-brain barrier is disrupted, can become a major determinant of tissue perfusion[12, 20]. Conversely, primary disturbances of the cerebral blood flow are an important factor of the development of vasogenic and cytotoxic brain oedema[7, 15, 18, 19, 40]. Analysis of the mutual relationship between blood flow disturbances and brain oedema formation, thus, is appropriate to facilitate evaluation of its clinical significance. Understanding of underlying interactions not only is of pathophysiological interest, but also pertinent in intensive care medicine as a basis of the clinical management and the development of more effective treatment. Formation of brain oedema and the development of cerebral blood flow disturbances are important manifestations of secondary brain damage in severe head injury and cerebral infarction[3], rendering these processes a central target for prevention and treatment.

Pathophysiology of Vasogenic Brain Oedema

Definitions of brain oedema appear appropriate in this context. The concept advanced by Klatzo[25] to distinguish brain oedema according to its vasogenic and cytotoxic nature has enhanced understanding of this complex phenomenon, although these prototypes occur simultaneously under clinical conditions, as in severe head injury, cerebral ischaemia or other lesions[2]. A simple albeit relevant definition characterizes brain oedema as an increase of the tissue water content leading to an increase of tissue volume[32]. The following discussion is focusing on vasogenic brain oedema, for example developing from cerebral contusion in head injury, or from an ischaemic necrosis as in cerebral infarction. Acute cerebral lesions with irreversibly damaged brain tissue are unavoidably associated with an opening of the blood-brain barrier[2, 19]. As the brain parenchyma proper, the cerebrovascular endothelium forming the blood-brain barrier is subjected to the processes causing tissue destruction. Consequently, vascular permeability is dramatically increased at the borderline of the injured tissue, not only for low-molecular weight solutes, electrolytes and water, but also for high-molecular weight components as plasma proteins. On the other hand, cellular components of the vascular compartment, such as red blood cells or leukocytes remain confined to the intravascular space, unless disruption of blood vessels has occurred as in a contusion focus[1].

The vasogenic oedema fluid which is emigrating from the irreversibly injured focus is spreading through extracellular routes, causing expansion of the interstitial fluid space and, thus, an increase of the regional tissue pressure in proportion with proximity to the focus[34]. The interstitial pressure rise can be considered as the major mechanism which compromises the regional microcirculation. Other pathophysiologically pertinent mechanisms, however, must be considered in addition, such as loss of cerebrovascular autoregulation and development of tissue acidosis, among others[11, 14]. Extravasation of brain oedema fluid under acute circumstances not only produces a space-occupying mass which may cause intracranial hypertension, but also affects the normal physiological composition of the interstitial fluid compartment. Vasogenic oedema fluid has a high protein content, mostly plasma proteins, which were extravasating through the leaky barrier[2, 33, 35]. Moreover, pathophysiologically active mediator compounds are accumulating in the interstitial oedema fluid, for example neurotoxic amino acids like glutamate or aspartate, free fatty acids including arachidonic acid, together with abnormal changes of the electrolyte composition, e.g., involving Ca^{2+}- and K^+-ions[2, 4, 5, 13]. These aspects notwithstanding, a major consequence of the oedematous fluid expansion is that maintenance of the regional cerebral blood flow becomes compromised. The adverse sequelae of vasogenic brain oedema can be summarized as follows:

- pathological alteration of interstitial fluid composition in the oedematous parenchyma with loss of extracellular fluid homeostasis
- release and accumulation of toxic mediator compounds, leading to secondary cell swelling and/or irreversible cell damage
- impairment of the regional cerebral microcirculation
- intracranial mass eventually causing intracranial hypertension once the intracranial compliance fails.

Cerebral Blood Flow in Acute Vasogenic Brain Oedema

Experimental and clinical evidence abounds on pathophysiological interactions between the formation of brain oedema and its adverse sequelae on the cerebral blood flow and O_2-supply of the brain. Frei *et al.*[12] were analyzing regional cerebral blood flow by the ^{85}Krypton clearance technique in animals with a focal cerebral lesion. rCBF measurements were made directly above and adjacent to the lesion as well as in remote areas or in the contralateral hemisphere at 24 h after induction of a cold injury of the cerebral cortex. A gradient of blood flow impairment was observed, from severe close to the lesion to moderate at distant sites or in the contralateral hemisphere[12]. Accumulation of oedema fluid was identified as the mechanism of the blood flow disturbances, which was particularly intensive in and adjacent to the lesion. Further, an increase of tissue lactic acid was found which appeared to follow the pattern of the regional oedema fluid accumulation under these conditions. The relationship between the gradient of blood flow disturbances and the gradient of oedema fluid accumulation was highly suggestive for an increased regional tissue pressure underlying the resulting flow deterioration.

This concept was tested on an experimental basis by measurements of regional tissue pressure of the brain after induction of vasogenic brain oedema by cold injury[34]. The pathological increase of tissue pressure was mostly pronounced in or adjacent to the lesion, while the pressure rise was mild at distant tissue areas, demonstrating generation of a pressure gradient from the lesion to peripheral sites. Build-up of the tissue pressure was a time-dependent process as confirmed during a posttraumatic observation period of up to 300 min. While under control conditions the tissue pressure was at the level of atmospheric pressure, i.e., 0 mmHg, it was increasing after trauma to 15 mmHg within 4–5 h in the vicinity of the lesion[34]. This observation formed the basis of an interesting pathophysiological concept, explaining accumulation and subsequent dissipation of vasogenic oedema fluid along a hydrostatic pressure gradient in the tissue as mechanism of oedema resolution by drainage into the ventricular system or subarachnoid space as clearance pathway[36].

A close relationship between oedema fluid accumulation and increase of tissue pressure was also reported in focal cerebral ischaemia[20]. The authors studied experimental animals with permanent occlusion of a middle cerebral artery. The tissue pressure was examined together with the ventricular fluid pressure, regional blood flow (H$_2$-clearance), and the regional tissue water content. A close correlation was obtained thereby between formation of ischaemic brain oedema in the caudate nucleus and a continuously increasing tissue pressure during a period of up to 4 h after MCA-occlusion[20]. The pressure gradient between the ischaemic focus and distant brain with normal flow eventually reached 8 mmHg. The elegant experiments demonstrate again an association between formation of brain oedema, the resulting tissue pressure increase, and its fatal consequences for the regional tissue perfusion. An intriguing aspect specific for focal cerebral ischaemia is that the resulting brain oedema is both, consequence and cause of primary and secondary disturbances of the cerebral blood flow.

Our laboratory has recently conducted investigations of regional blood flow changes in the brain following a focal lesion with extravasation of vasogenic oedema for assessment of effects thereon of frequently used anaesthetic agents, such as fentanyl, isoflurane, or thiopental[9, 29, 30]. The underlying question was, whether and how these anaesthetic methods affect formation of vasogenic brain oedema from an acute focal lesion with regard to respective blood flow alterations. The experiments correspond to the clinical situation of induction of anaesthesia for diagnostic or surgical procedures in patients with

an acute cerebral lesion, for example from head injury. In a subgroup of experimental animals, the focal lesion (cold injury) was combined with a transient period of systemic hypoxia by hypoventilation, as is occurring in emergency conditions in patients. The regional water content of the brain was measured in multiple samples of cerebral cortex close to and distant from the lesion, as well as in specimen taken from the contralateral hemisphere.

The water content of oedematous tissue samples studied by the specific gravity method was higher in animals with thiopental anaesthesia than in animals with isoflurane[9]. Parallel measurements of the regional cerebral blood flow (H$_2$-clearance) close to the lesion and at distant areas during up to 6 h after trauma demonstrated an early hyperaemic response irrespectively of the type of anaesthesia. Acute hyperaemia was followed then by a decrease of blood flow to subnormal levels in animals with fentanyl, while hyperaemia persisted in animals with isoflurane. In animals with thiopental anaesthesia, the blood flow response to the lesion was comparatively mild as compared to animals with isoflurane or fentanyl, respectively[30]. The variety of well controlled studies indicate altogether that development of vasogenic brain oedema from a focal lesion immediately as well as 24 h later – the time period of the maximum of oedema accumulation – is associated with marked alterations of regional blood flow, among others attributable to an increased tissue water content and brain tissue pressure.

Ischemic Brain Oedema

The following part of this analysis is dealing with the brain oedema – blood flow relationship from an inverse point of view, namely whether and how primary disturbances of cerebral blood flow are causing brain oedema. This pathophysiological problem has been analyzed by studying both, focal as well as global forms of blood flow disturbances. Ischaemic flow thresholds could be identified thereby which were associated with an increase of the tissue water content, once blood flow was falling to or below a critical level. Evidence so far accumulated is in agreement that decreasing cerebral tissue perfusion below 40–30% of normal raises cerebral water content. Such flow thresholds were identified in studies on focal cerebral ischaemia in combination with measurements of the regional tissue water content[40].

The pioneering studies of Symon and coworkers[7, 40] of this aspect were confirmed by investigations of Hossmann et al.[19] or Iannotti et al.[20] among others, who observed formation of ischaemic brain oedema, once the regional tissue perfusion was falling below 10–20 ml/100 g × min. It is noteworthy that such a flow level is in the neighbourhood of the flow range which is associated with the formation of irreversible damage of brain parenchyma[21]. Accordingly, reduction of blood flow below ca. 20% normal causes irreversible brain damage, if the tissue perfusion is not normalized within a few hours. It might, therefore, be concluded that formation of ischaemic brain oedema is related with the irreversible destruction of brain tissue by ischaemia.

Improvements in the technology of measuring blood flow and other functional parameters at high spatial resolution using autoradiography made possible the discovery of ischaemic flow thresholds, which appear to be selectively associated with distinct functional measures, such as energy metabolism or protein synthesis. Mies et al.[28] have demonstrated inhibition of protein synthesis and onset of energy failure as reflected by ATP-depletion in the brain of animals with MCA-occlusion to occur at different ischaemic flow thresholds. The studies, however, made equally obvious the significance of the duration of the blood flow disturbances. Whereas inhibition of protein synthesis became noticeable almost immediately after vessel occlusion and at relatively high flow levels, the energy state was initially maintained, but only started to break down with considerable delay. At 12 h after vessel occlusion the ischaemic flow thresholds associated with inhibition of protein synthesis and induction of energy failure were approaching each other[28].

An important aspect of ischaemic brain oedema is the specific nature of the ischaemic insult namely focal or global cerebral ischaemia. Permanent occlusion of a major blood vessel, such as the middle cerebral artery in experimental animals is followed during subsequent days by a continuous increase of the cerebral water content of the affected brain parenchyma, reaching a maximum at 3 days post occlusion[15]. The brain water content then decreases during the following days. Accumulation of oedema fluid in the ischaemic hemisphere is paralleled by respective changes of the Na^+-content, whereas the K^+-content of the tissue is decreasing in an almost mirror-like fashion, reaching a minimum in about 3 days[15]. It is noteworthy that the early increase of tissue water content after MCA occlusion appears to be independent from gross disruption of the blood-brain barrier. This was concluded from measurements of the uptake of radioactively labelled serum albumin into the ischaemic oedematous hemisphere. Contrary to the early increase of the tissue water content, accumulation of the macromolecular blood-brain barrier marker was delayed for 2 to 3 days[15]. This finding supports the contention that ischaemic brain-oedema evolving immediately after vessel occlusion is of cytotoxic nature, whereas development of the vasogenic component follows with a delay of days. It implies that therapeutic measures against brain oedema, such as hypertonic solutions should be more effective early after vessel occlusion, as long as the blood-brain barrier is intact.

The development of brain oedema from global interruptions of cerebral blood flow has characteristic features which distinguish this entity from brain oedema following focal cerebral ischaemia. Experiments of Betz[10] and Hossmann[18] have unanimously shown that during the period of complete interruption of blood flow to the brain, the brain bulk or water content, respectively remains at a normal level, as does the intracranial pressure. Nevertheless, the ischaemic period is characterized by marked changes of brain tissue pH, shrinking of the extracellular fluid compartment from translocation of fluid into the intracellular space, and increase of tissue osmolality[10, 18, 41]. However, once blood flow is re-established, an almost explosive increase of the cerebral water content follows together with the development of intracranial hypertension[18]. The further course of brain oedema is dependent then on the quality of the postischaemic reperfusion of the brain. It may eventually resolve provided the cerebral tissue is adequately perfused, whereas a low or subnormal blood flow further enhances ischaemic brain oedema and, thereby, ruins the chances for recovery of the brain[17].

Disruption of Blood-Brain Barrier in Cerebral Ischaemia

Although it is well accepted that the quality of postischaemic reperfusion determines whether or not cerebral function returns, evidence is available that early development of brain oedema, with opening of the blood-brain barrier may depend on the level of postischaemic hyperaemia. Seida et al.[37] studied in experimental animals with 1 h of middle cerebral artery occlusion, whether attenuation of the reactive hyperaemic response affects development of brain oedema after re-opening of the arterial blood vessel. rCBF was measured by H_2-clearance, tissue water content by specific gravity at 3 h of postischaemic recirculation. Mitigation of reactive hyperaemia was attempted by infusion of aminophylline. Untreated animals with 1 h of MCA-occlusion had an approximately 5-fold increase of blood flow in the ipsilateral caudate nucleus 10 min after opening of the blood vessel, whereas in animals with aminophylline postischaemic hyperaemia was limited to a 2-fold increase above control[37]. Most importantly, gross opening of the blood-brain barrier in the caudate nucleus or cerebral cortex was observed in 4 of 5 animals without treatment, whereas only in 1 animal with aminophylline. Protection of the blood-brain barrier by attenuation of reactive hyperaemia seemed also to inhibit the development of ischaemic tissue damage. This was inferred from histological investigations of the brain conducted at 3 days or 2 weeks after the insult[37].

The pathophysiological significance of reactive hyperaemia was confirmed in studies on focal cerebral ischaemia by MCA-occlusion of 3 h duration followed by postischaemic reperfusion[26]. The level of postischaemic hyperaemia was controlled in these investigations by mechanical constriction of the middle cerebral artery after release of vessel occlusion. The flow obstruction also inhibited reactive hyperaemia and thereby opening of the blood-brain barrier with development of postischaemic brain oedema[26]. The pathological role of uncontrolled reactive hyperaemia notwithstanding, it is well established that failure of an adequate reperfusion of the brain after ischaemia is of equal or even worse significance for cerebral recovery[17]. The apparent consequences for treatment of cerebral ischaemia and ischaemic brain oedema are intriguing. Hence, efforts must be made on the one hand to re-establish a normal cerebral perfusion pressure – a formidable requirement after general circulatory failure – while on the other hand development of an uncontrolled and overwhelming hyperperfusion must be avoided under all circumstances once the general circulation has resumed function.

Ischaemic Cell Swelling – Cytotoxic Brain Oedema

Some final comments on the significance of cell swelling, the manifestation of cytotoxic brain oedema, are appropriate. As discussed, early formation of brain oedema after interruption of cerebral blood flow concerns the development of cytotoxic swelling of nerve- and glial cell elements from uptake of extracellular fluid[2, 18, 41]. As to underlying mechanisms the interpretation is that energy failure resulting from ischaemia is responsible, causing breakdown of the extra/intracellular Na^+-concentration gradient and, hence, accumulation of water in the cells[27]. This conventional view, however, might be too narrow. By employment of a well controlled in-vitro system using cell lines of central nervous origin for studying conditions occurring in ischaemia cell swelling was found to evolve, e.g. by acidosis, excitatory amino acids, high K^+-levels in the medium, or after addition of toxic lipids, such as arachidonic acid[6, 24, 39]. These in vitro studies were conducted in the presence of an ample O_2- and substrate supply for the cells, ruling out a role of energy failure in the swelling process. The independence of cell swelling from breakdown of energy metabolism, from depletion of metabolic fuel for the ion pumps, is underscored by observations that exposure of viable cells (C6 glioma) in a completely anoxic environment with inhibition of glycolysis neither induces cell swelling nor impairs cell viability[23].

It is nevertheless obvious that re-establishment of a normal cell volume after swelling requires a functional energy metabolism. This was shown by in vitro investigations on osmotic cell swelling with and without inhibition of the energy metabolism[23, 31]. Briefly, cell volume of glial cells suspended in a hypotonic medium is rapidly increasing which, however, is followed by spontaneous recovery of the cells to normal size, a phenomenon known as regulatory volume decrease[22]. If, however, osmotic cell swelling is induced in complete anoxia together with inhibition of glycolysis by iodoacetate, the regulatory cell volume decrease after osmotic swelling is largely inhibited[23, 31].

Initial cell swelling occurring in ischaemic brain tissue in vivo, therefore, is unlikely to be the result of a failure of basic volume regulatory processes, but on the contrary is most probably due to the control mechanisms working to re-establish homeostasis as a requirement of normal function. Experimental evidence supporting this contention is available. As frequently shown ischaemia is associated with a marked increase of extracellular K^+-levels, release and accumulation of neuroexcitatory transmitter compounds, such as glutamate, and development of acidosis, among others[5, 8, 10, 19]. The resulting cell swelling, involving glial cells in particular, might be viewed as a salvage response to clear interstitial fluid of abnormal K^+-concentrations, transmitter concentrations, and other material[6]. Due to the uphill concentration gradient of K^+ or glutamate from the extra- to the intracellular compartment, their transport must involve active uptake mechanisms. The cell volume increase from intracellular acidosis can also be viewed as a compensatory response designated to eliminate excessively accumulating H^+-ions from the intracellular compartment in order to normalize intracellular pH[16, 24, 38, 39]. The active transport of e.g. glutamate into cells is fuelled by a concurrent influx of Na^+-ions from the extracellular compartment. Actually, energy for clearance of glutamate from the extracellular space by the glial cells is provided by the simultaneous downhill influx of Na^+-ions, which together with the amino acid are raising cell osmolality. Conversely, removal of H^+-ions by the Na^+/H^+-antiporter from the cell is accomplished by a stochiometric exchange against extracellular Na^+-ions[16]. This again may result in accumulation of Na^+-ions together with water in the cells as the final mechanism of acidosis induced cell swelling[24, 39].

Thus, cell swelling occurring early in cerebral ischaemia does not necessarily indicate cell damage, but rather activation of regulatory processes to maintain or re-establish, respectively physiological conditions required for normal nerve cell function. Cell swelling as the result of such compensatory mechanisms, however has a price, namely an increase of brain tissue mass which occupies additional intracranial space with the potential of causing intracranial hypertension. Nevertheless, therapeutic methods employed to inhibit cell swelling due to ischaemia or other pathological conditions must not interfere with the physiological purpose of the swelling process, namely to normalize tissue homeostasis in the brain, which already is in jeopardy from trauma or ischaemia.

Acknowledgement

The secretarial assistance for the preparation of the manuscript of Helga Kleylein and Monika Stucky is gratefully acknowledged.

References

1. Adams JH, Graham DI, Gennarelli TA (1986) Primary brain damage in non-missile head injury In: Baethmann A, Go KG, Unterberg A (eds) Mechanisms of secondary brain damage. NATO ASI Series, Plenum Press, New York, pp 1–13 (Life Sciences, Vol 115)
2. Baethmann A (1978) Pathophysiological and pathochemical aspects of cerebral edema. Neurosurg Rev 1: 85–100
3. Baethmann A, Go KG, Unterberg A (1986) Mechanisms of secondary brain damage. NATO ASI Series A. Plenum Press, New York (Life Sciences, Vol 115)
4. Baethmann A, Maier-Hauff K, Kempski O, Unterberg A, Wahl M, Schürer L (1988) Mediators of brain edema and secondary brain damage. Crit Care Med 16: 972–978
5. Baethmann A, Maier-Hauff K, Schürer L, Lange M, Guggenbichler Ch, Vogt W, Jacob K, Kempski O (1989) Release of glutamate and of free fatty acids in vasogenic brain edema. J Neurosurg 70: 758–591
6. Baethmann A, Kempski O (1992) Biochemical factors and mechanisms of secondary brain damage in cerebral ischemia and trauma In: Bazan NG, Braquet P, Ginsberg D (eds) Neurochemical correlates of cerebral ischemia, Vol 7. Advances in Neurochemistry, Plenum Press, New York, pp 295–320
7. Bell BA, Symon L, Branston NM (1985) CBF and time thresholds for the formation of ischemic cerebral edema, and effect of reperfusion in baboons. J Neurosurg 62: 31–41
8. Benveniste H, Drejer J, Schousboe A, Diemer NH (1984) Elevation of the extracellular concentrations of glutamate and aspartate in rat hippocampus during transient cerebral ischemia monitored by intracerebral microdialysis. J Neurochem 43: 1369–1374
9. Berger S, Murr R, Schürer L, Baethmann A (1993) The development of brain edema after a focal cerebral lesion: influence of anesthesia and of hypoxia (in preparation)
10. Betz E (1977) Vascular reactivity and ion homeostasis in heart and brain In: Zülch KJ, Kaufmann W, Hossmann K-A, Hossmann V (eds) Brain and heart infarct. Springer, Berlin Heidelberg New York, pp 10–18
11. Dittmann J, Herrmann H-D, Loew F (1972) Examination of the metabolism of oedematous brain tissue. Acta Neurochir (Wien) 27: 63–85
12. Frei HJ, Wallenfang Th, Pöll W, Reulen HJ, Schubert R, Brock M (1973) Regional cerebral blood flow and regional metabolism in cold induced oedema. Acta Neurochir (Wien) 29: 15–28
13. Go KG (1986) Disturbances of extracellular homeostasis after a primary insult as a mechanism in secondary brain damage In: Baethmann A, Go KG, Unterberg A (eds) Mechanisms of secondary brain damage. NATO ASI Series, Plenum Press, New York, pp 127–137 (Life Sciences, Vol 115)
14. Go KG, Zijlstra WG, Flanderijn H, Zuiderveen F (1974) Circulatory factors influencing exudation in cold-induced cerebral edema. Exp Neurol 42: 332–338
15. Gotoh O, Asano T, Koide T, Takakura K (1985) Ischemic brain edema following occlusion of the middle cerebral artery in the rat. I: The time courses of the brain water, sodium and

potassium contents and blood-brain barrier permeability to ^{125}I-albumin. Stroke 16: 101–109

16. Grinstein S, Rothstein A (1986) Mechanisms of regulation of the Na^+/H^+-exchanger. J Membr Biol 90: 1–12

17. Hossmann KA, Lechtape-Grüter H, Hossmann V (1973) The role of cerebral blood flow for the recovery of the brain after prolonged ischemia. Z Neurol 204: 281–299

18. Hossmann KA (1976) Development and resolution of ischemic brain swelling In: Pappius HM, Feindel W (eds) Dynamics of brain edema. Springer, Berlin Heidelberg New York, pp 219–227

19. Hossmann K-A (1984) Pathophysiologie der Hirndurchblutung. In: Paal G (ed) Therapie der Hirndurchblutungsstörungen. Edition Medizin, Weinheim, pp 37–84

20. Iannotti F, Hoff JT, Schielke GP (1985) Brain tissue pressure in focal cerebral ischemia. J Neurosurg 62: 83–89

21. Jones TH, Morawetz RB, Crowell RM, Marcoux FW, Fitzgibbon SJ, DeGirolami U, Ojemann RG (1981) Thresholds of focal cerebral ischemia in awake monkeys. J Neurosurg 54: 773–782

22. Kempski O, Chaussy L, Groß U, Zimmer M, Baethmann A (1983) Volume regulation and metabolism of suspended C6 glioma cells: an in vitro model to study cytotoxic brain edema. Brain Res 279: 217–228

23. Kempski O, Zimmer M, Neu A, v. Rosen F, Jansen M, Baethmann A (1987) Control of glial cell volume in anoxia – in vitro studies on ischemic cell swelling. Stroke 18: 623–628

24. Kempski O, Staub F, Jansen M, Schödel F, Baethmann A (1988) Glial swelling during extracellular acidosis in vitro. Stroke 19: 385–392

25. Klatzo I (1967) Presidential address – Neuropathological aspects of brain edema. J Neuropath Exp Neurol 26: 1–14

26. Kuroiwa T, Shibutani M Okeda R (1989) Nonhyperemic blood flow restoration and brain edema in experimental focal cerebral ischemia. J Neurosurg 70: 73–80

27. Macknight ADC, Leaf A (1977) Regulation of cellular volume. Physiol Rev 37: 510–573

28. Mies G, Ishimaru S, Xie Y Seo K, Hossmann K-A (1991) Ischemic thresholds of cerebral protein synthesis and energy state following middle cerebral artery occlusion in rat. J Cereb Blood Flow Metab 11: 753–761

29. Murr R, Schürer L, Berger S, Enzenbach R, Baethmann A (1991) Effects of anesthetic agents on brain edema and cerebral blood flow from a focal lesion in rabbit brain. Adv Neurosurg 19: 24–28

30. Murr R, Schürer L, Berger S, Enzenbach R, Peter K, Baethmann A (1993) Effects of isoflurane, fentanyl, or thiopental anesthesia on regional cerebral blood flow and brain surface PO_2 in the presence of a focal lesion in rabbits. Anesth Analg (in press)

31. Neu AH (1990) In-vitro Untersuchungen zum zytotoxischen Hirnödem: Regulation des Zellvolumens hypoton suspendierter C6-Gliomzellen mit zusätzlicher Stoffwechselinhibition. Thesis, Ludwig-Maximilians-University, München

32. Pappius HM (1974) Fundamental aspects of brain edema. In: Vinken PJ, Bruyn GW (eds) Part I: Tumors of the brain and skull. Handbook of clinical neurology, Vol 16. North Holland Publ., Amsterdam American Elsevier, New York, pp 167–185

33. Ramussen LE, Klatzo I (1969) Protein and enzyme changes in cold injury edema. Acta Neuropathol (Berl) 13: 12–28

34. Reulen HJ, Kreysch HG (1973) Measurement of brain tissue pressure in cold induced cerebral oedema. Acta Neurochir (Wien) 29: 29–40

35. Reulen HJ (1976) Vasogenic brain edema. New aspects in its formation, resolution and therapy. Br J Anesth 48: 741–751

36. Reulen HJ, Tsuyumu M, Tack A, Fenske A, Prioleau G (1978) Clearance of edema fluid into cerebrospinal fluid. A mechanism for resolution of vasogenic brain edema. J Neurosurg 48: 754–764

37. Seida M, Wagner HG, Vass K, Klatzo I (1988) Effect of aminophylline on postischemic edema and brain damage in cats. Stroke 19: 1275–1282

38. Siesjö BK (1985) Acid-base homeostasis in the brain: physiology, chemistry, and neurochemical pathology In: Kogure K, Hossmann K-A, Siesjö BK, Welsh FA (eds) Molecular mechanisms of ischemic brain damage. Elsevier, Amsterdam, pp 121–154 (Progress in brain research, Vol 63)

39. Staub F, Baethmann A, Peters J, Weigt H, Kempski O (1990) Effects of lactacidosis on glial cell volume and viability. J Cereb Blood Flow Metab 10: 866–876

40. Symon L (1986) Progression and irreversibility in brain ischaemia. In: Baethmann A, Go KG, Unterberg A (eds) Mechanisms of secondary brain damage. NATO ASI Series, Plenum Press, New York, pp 221–238 (Life Sciences, Vol 115)

41. Van Harreveld A (1972) The extracellular space in the vertebrate central nervous system In: Bourne GH (ed) The structure and function of nervous tissue, Vol 4. Academic Press, London, pp 447–511

Correspondence and Reprints: Dr. Reinhart Murr, Institute of Anaesthesiology, Klinikum Großhadern, Ludwig-Maximilians-University, Marchioninistrasse 15, D-81377 Munich, Federal Republic of Germany.

Acta Neurochir (1993) [Suppl] 59: 18–21

Measurement of Vascular Reactivity in Head Injured Patients

A. Marmarou, K. Bandoh, M. Yoshihara, and **O. Tsuji**

Division of Neurosurgery, Medical College of Virginia, Richmond, Virginia, U.S.A.

Summary

It is has been demonstrated that clinical outcome following head injury is correlated with the reactivity of the cerebrovasculature to carbon dioxide changes. Since CBF measurements are difficult to perform in these patients, a new technique is proposed utilizing the ICP response to capnic stimuli.

In 40 head injured patients, the responses of ICP, pressure volume index (PVI) and middle cerebral artery velocities to hypocapnia and to hypercapnia were determined.

Hypocapnia reduced ICP and MCA velocity while hypercapnia was followed by ICP and MCA velocity increases. Both changes were in the same magnitude supporting the concept the global ICP response reflects vascular reactivity.

The fact that the velocity response to hypocapnia in lesioned hemispheres was less compared to the ICP response indicates the loss of ability to dilate in injured vessels and is consistent with earlier findings relating reduced reactivity to poor outcome.

Keywords: Head injury; vascular reactivity; ICP; PVI; hypocapnia; hypercapnia; CBF velocity; outcome.

Introduction

The ability of the cerebrovasculature to alter blood flow (CBF) and volume by change in vessel caliber is essential for maintaining stability in perfusion and reducing intracranial pressure (ICP). In the clinical setting, this reactivity has been defined in terms of CBF change to alterations in arterial carbon dioxide tension. ($PaCO_2$) and clinical reports have shown that reactivity is correlated to clinical outcome[1-5]. As a result, serial CBF measures requiring relatively complex equipment have been necessary to study alterations in reactivity in the brain injured patient. Recently, a new technique was introduced which utilized the ICP response to capnic stimuli as the basis for developing an indirect method of vascular reactivity assessment[6, 7]. In part, the development of this technique was prompted by our findings that CSF parameters played a minor role in the development of intracranial pressure. Thus, we examined means by which the cerebrovasculature could be challenged to permit a more quantitative assessment of the blood volume/intracranial pressure relationship and hopefully to derive a suitable reactivity measure. Clearly, an assessment of vascular/pressure interaction would have to be non-invasive and practical in the clinical setting. This report provides a brief summary of our efforts in developing this method and presents our preliminary results in applying this technique to the severely head injured patient.

Methodology

Definition of Reactivity Indices

The basis for the development of the pressure and volume reactivity indices resides in the intracranial pressure volume curve. We considered the baseline ICP as one stable point on the curve which is normally challenged by altering CSF volume. However, if the assumption is made that CSF volume remains fixed during change in $PaCO_2$, the resulting change in ICP can be attributed to change in blood volume in response to vasoconstriction or vasodilation. This allows the definition of indices to capnic stimulation to be expressed either in terms of pressure or utilizing the PVI in terms of volume. Firstly, the pressure indices were defined simply as the change in ICP in response to a change in end-tidal carbon dioxide tension ($PeCO_2$) or $\Delta ICP/\Delta PeCO_2$. The term PRCO refers to the pressure response to hypercapnia produced by mild hypoventilation and PRCR for hypocapnia produced by mild hyperventilation. In terms of volume, the estimated blood volume change (ΔEBV) for a change in $PeCO_2$ may be calculated from the equation $\Delta EBV = PVI \times [Log(Po/Pe)]$ where PVI is the pressure volume index measured just prior to the study, Po the baseline pressure and Pe the level of ICP reached in response to the capnic stimulation. Having estimated the change in blood volume, the blood volume responsivity (BVR) to capnic stimulation is simply given by ($\Delta EBV/\Delta PeCO_2$). Using this equation BVRo and BVRr defined the blood volume responsivity to hypercapnia and hypocapnia respectively. The subscripts in these terms were selected to correspond to the method by which they were determined. For example, BVRo was obtained by inducing hypoventilation and BVRr was obtained in response to hyperventilation. All results are reported as mean and plus or minus one standard deviation unless otherwise noted.

Study Population

The study population consisted of severely head injured patients (n = 40, GCS < 8), 30 males and 10 females of median age 32.7 ± 17.2 years. The average coma score of this cohort equaled 6.0 ± 1.4.

Primary diagnosis included 33 closed head injuries and 7 skull fractures. The CT findings of the 33 patients with closed head injury included 9 with normal CT, 11 acute subdural hematoma, 6 focal contusion, 1 epidural hematoma, 5 brain stem injury, 1 diffuse axonal injury and 9 with normal CT scan.

Study Protocol

The study protocol was approved by an internal review board and consent was obtained from families of those patients participating in this study. The first study was usually performed within 24 hours post admission and was continued daily for 5 days. Prior to each study, the PVI was determined from bolus technique. The MCV treatment protocol for ICP is standardized and consisted of a five level scale administered in staircase fashion as required to maintain ICP below 20 mmHg. In brief: level 1 sedation, level 2 drainage, level 3 Mannitol, level 4 hyperventilation, and level 5 barbiturates. Patients with high ICP were excluded from the hypercapnic study protocols.

Transcranial Doppler (TCD) Measures

Measurement of left and right middle cerebral artery (MCA) velocities were obtained daily using a TC2-64 transcranial doppler instrument (Eden Medical Electronics Inc., Koustenz, Germany). The reactivity in terms of velocity followed closely the terminology and protocol for assessment of pressure and blood volume indices. The change in MCA velocity to hypocapnia was designated as VRCO and to hypercapnia VRCR. Usually, the change in $PeCO_2$ averaged 2 to 3 torr above and below the baseline level.

Results

Temporal Course: Physiologic Parameters

ICP and CPP

The ICP averaged 13.8 ± 5.1 s.d. on day 1 and gradually increased to a maximum daily average of 17.9 ± 4.9 s.d. and 16.5 ± 4.0 s.d. on days 4 and 5 respectively. Average daily cerebral perfusion pressure was 83.2 ± 16.3 s.d. on day 1 and reached a minimum value of 76.1 ± 19.0 s.d. by day 4.

Pressure Volume Index (PVI)

The PVI measured within the first 24 hours averaged 24.7 ± 7.6 ml and gradually decreased to a

level of 20.2 ± 7.7 ml by day 5. PVI values over the 5 day period were within this range.

End Tidal $PeCO_2$

The $PeCO_2$ averaged 29.3 ± 4.5 mmHg on day 1 in response to moderate hyperventilation. The daily average of $PeCO_2$ remained relatively constant and the lowest average value measured on day 5 was 27.8 ± 3.9 s.d. mmHg. Average values for all days fell within this range.

Transcranial Doppler: MCA Velocities

The MCA velocities were classified into two groups according to CT criteria. Group 1 included those hemispheres with contusion, intracerebral hematoma or diffuse swelling and were designated the "lesion" side. Group 2 included those brain hemispheres exhibiting no pathology by CT and were designated the "non-lesion" side.

The MCA velocities were within normal ranges on day 1 and averaged 52.0 ± 12.0 cm/sec and 55.6 ± 12.1 cm/sec on non lesion and lesion sides respectively. However, as time progressed, MCA velocities on non-lesion and lesion hemispheres gradually increased, reaching a peak of 72.2 ± 27.7 cm/sec and 82.7 ± 20.8 by day 4 before declining to 69.7 ± 24.9 and 76.9 ± 22.1 on day 5 respectively. This increase over baseline values reached statistical significance by day 2 (p < 0.01). The MCA velocity in lesioned hemispheres was consistently lower than in non-lesioned hemispheres although these differences did not reach statistical significance. In three patients, MCA velocities exceeded 100 cm/sec approaching levels associated with vasospasm. Pulsatility indices over the five day course were very consistent and ranged from a high of 0.96 ± 0.22 on day 1 and a low of 0.81 ± 0.18 measured on day 2.

Pressure Reactivity (PRCO, PRCR)

The pressure reduction achieved per torr reduction of $PeCO_2$ achieved by hyperventilation averaged 0.6 ± 0.6 on day 1 and was consistently below the response to hypercapnia which averaged 1.4 ± 1.4 for the same time period. Of the 266 studies performed in the 40 patients over the 5 day period, an ICP response to capnic stimulation was obtained in 238 studies (89%). Pressure responsivity was absent in only 2 of 24 patients during the first 24 hours post admission.

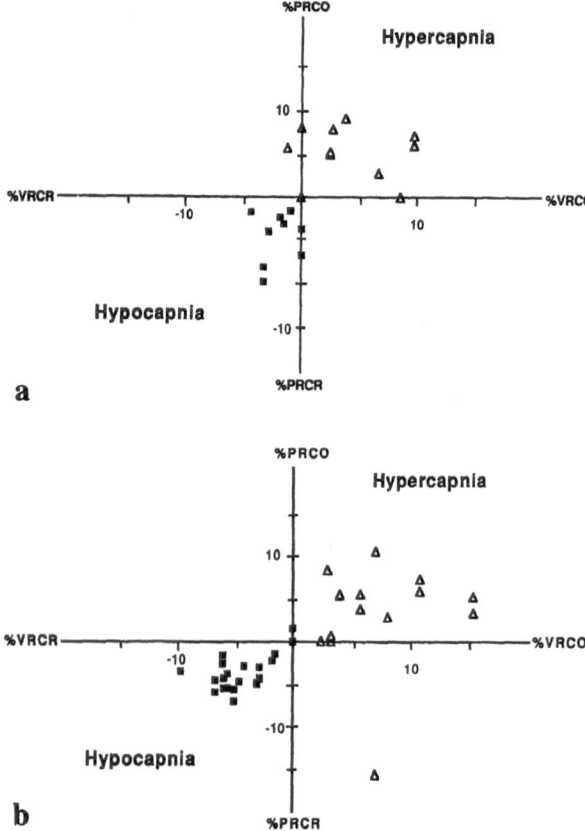

Fig. 1. ICP and MCA responses to hypercapnia and hypocapnia in lesion (a) and non-lesion hemispheres (b) as observed on CT. A good agreement was observed between the direction of change in global ICP response and flow velocity. Reactivity to hypocapnia, or the ability of the vessels to dilate was less in hemispheres with lesion

Agreement of ICP and MCA responses to capnic stimulation

The ICP and MCA response indices obtained during these studies were compared for lesion and non-lesioned hemispheres (cf. Fig. 1). For the non-lesion hemisphere, changes in ICP in response to capnic stimulation were accompanied by changes in MCA velocity in the majority of cases (31/34, 88%). However, for hemispheres with lesion, the agreement was less (15/20, 75%) and the response to hypocapnia was blunted. When these indices were compared according to GOS, patients with favorable (good/moderate) outcome had 100% agreement while patients with unfavorable outcome (sev/veg/dead) had only a 68.5% agreement.

Relationship of ICP and Blood Volume

For a median PVI level of 21, the average change in blood volume per torr increase in $PeCO_2$ equaled

$3.93 \, ml \pm 0.55$ for hypercapnia and $-1.97 \pm 0.28 \, ml$ for hypocapnia. Dividing the estimated blood volume by the change in ICP, it was found that only $0.5 \, ml$ of blood volume change was necessary with hypercapnia to produce a change of $1 \, mmHg$ ICP while a slightly greater amount ($0.56 \, ml$) was required with hypocapnia.

Discussion

These studies demonstrate that the pressure responses to capnic stimulation are quite consistent and easily determined in a clinical setting. Heretofore, the relationship of ICP and endtidal $PeCO_2$, had not been studied in head injured patients. The observation that changes in ICP are in agreement with changes in MCA velocity in both direction and magnitude would support the concept that the global ICP response reflects vascular reactivity. More specifically, the ICP follows the blood volume change produced by the constriction and dilation of the resistance vessels as they respond to capnic stimulation. Interestingly, the state of the vessels can be extracted by the relative magnitude of PRCR which reflects the ability of the vessel to constrict, and PRCO which reflects the ability of the vessel to dilate. As PRCR values were consistently lower than PRCO, it would suggest that the microvasculature is in a state of sustained vasoconstriction or compression. The fact that MCA velocities were only moderately increased in the majority of patients would suggest that this compression is probably not due to vasospasm. Muizelaar's findings of reduced blood volume in areas of ischemia (personal communication) is consistent with these findings. Moreover, our observation that the ability of the vessels to dilate is reduced in the injured hemisphere with mass lesion is in general agreement with previous reports relating reduced reactivity to poor neurologic outcome.

Another interesting observation is that the amount of blood volume necessary to alter ICP is quite small. For example, for a baseline ICP of $15 \, mmHg$ and PVI of 21, these data demonstrate that only $2.5 \, ml$ change in blood volume would be necessary to reach critical ICP levels of $20 \, mmHg$. This volume would even be less as PVI is further reduced by swelling.

The ability of these measures to provide a reliable estimate of vascular reactivity depends upon two major assumptions. First, that the cerebrospinal fluid volume remains constant during capnic maneuvers. We pose that this assumption is reasonable for the short duration of these capnic challenges. A second assumption is that the larger arteries do not change caliber. If this were true, the blood volume reduction could not be attributed solely to the resistance vessels. Thus, it appears that the reliability of both ICP and transcranial doppler measures of reactivity depend upon the dimensional stability of the main arteries. Future studies will directly compare ICP, TCD and CBF reactivity to help resolve these issues.

References

1. Cold GE, Jensen FT, Malmros R (1977) The cerebrovascular CO_2 reactivity during the acute phase of brain injury. Acta Anaesthesiol Scand 21: 222–231
2. Enevoldsen EM, Jensen FT (1978) Autoregulation and CO_2 responses of cerebral blood flow in patients with acute severe head injury. J Neurosurg 48: 689–703
3. Overgaard J, Tweed WA (1974) Cerebral circulation after head injury. Part I: Cerebral blood flow and its regulation after closed head injury with emphasis on clinical correlations. J Neurosurg 41: 531–541
4. Schalén W, Messeter K, Nordström CH (1991) Cerebral vasoreactivity and the prediction of outcome in severe traumatic brain lesions. Acta Anaesthesiol Scand 35: 113–122
5. Tenjin H, Yamaki T, Nakagawa Y, et al (1990) Impairment of CO_2 reactivity in severe head injury patients: an investigation using thermal diffusion method. Acta Neurochir (Wien) 104: 121–125
6. Marmarou A, Wachi A (1988) Blood volume responsivity to ICP change in head injured patients. In: Miller JD, et al (eds) Intracranial Pressure VII. Springer, Berlin Heidelberg New York, pp 688–690
7. Marmarou A (1991) Increased intracranial pressure in head injury and influence of blood volume. J Neurotrauma 8 [Suppl 2]: 301–306

Correspondence and Reprints: Dr. A. Marmarou, Division of Neurosurgery, Medical College of Virginia, P. O. Box 508, MCV Station, Richmond, VA 23298, U.S.A.

Intermittent Measurements of Cerebral Blood Flow and Metabolism in Intensive Care

Acta Neurochir (1993) [Suppl] 59: 25–27

Measurements of Cerebral Blood Flow and Metabolism in Severe Head Injury Using the Kety-Schmidt Technique

C. Robertson

Department of Neurosurgery, Baylor College of Medicine, Houston, Texas, U.S.A.

Summary

Global cerebral blood flow (CBF) was measured serially for up to 10 days after severe head injury, and related to outcome. Twenty-five of the patients had a reduced CBF, 47 had a normal CBF, and 30 had an elevated CBF. Patients with a reduced CBF had a poorer outcome than patients with a normal or elevated CBF. There were no differences in the type of injury, initial GCS, severity of intracranial hypertension in each CBF group. Systemic factors did not significantly contribute to the differences in CBF among the 3 groups.

A logistic regression model of the effect of CBF on neurological outcome was developed. When adjusted of variables which were found to be significant confounders, including age, initial Glasgow Coma Score, hemoglobin, cerebral perfusion pressure, and cerebral oxygen consumption, a reduced CBF remained significantly associated with an unfavorable neurological outcome.

Keywords: Severe head injury; outcome; global CBF; Kety-Schmidt technique.

Introduction

Several previous studies have examined the relationship between CBF and neurological outcome after head injury. Both reductions and elevations of CBF have been associated with a poor neurological outcome.

Overgaard *et al.* reported regional CBF values of <0.2 ml/gm/min in 61% of patients with poor neurological outcomes. Only 10% of patients with a good outcome had regional CBF values of this low level[3]. Overgaard and Tweed reported that patients who died or remained vegetative commonly had very low CBF in arterial boundary zones and had lower global CBF values than patients who recovered[4]. Other studies have correlated a moderately reduced CBF, particularly in the first 24 hours after injury, with a poor outcome[3, 6].

Elevated CBF levels have also been related to a poor neurological outcome. Fieschi *et al.* reported that an initially low CBF returned to normal in patients who recovered but increased to elevated levels in patients who died of the head injury[1]. Overgaard and Tweed reported a good outcome in only 1 or 12 patients in their series with an elevated CBF[4].

The purpose of this study was to examine the relationship between serial measurements of global CBF and neurological outcome after severe head injury.

Methods

Patient Characteristics

A total of 102 adult patients admitted between 1983 and 1989 had repeated measurements of CBF for up to 10 days after injury. Details of these patients have been reported in a previous publication[5]. Eighty percent of the patients were male with a mean age of 34 ± 11 years. Ninety percent had closed head injuries and 10% had gunshot wounds of the head. The initial GCS was 3–5 in 21% of the patients and 6–8 in 73%. In 6% of the patients, the initial GCS was >8, but these patients later deteriorated to coma.

CBF Monitoring

Global CBF and cerebral metabolic rate of oxygen and lactate were measured at least daily, using the nitrous oxide saturation method, for as long as ICP monitoring was required. A total of 739 measurements of CBF (mean 7 per patient) were obtained.

The nitrous oxide saturation method for measuring global cerebral blood flow (CBF) was initially described in 1948 (Kety 1948). There are several advantages of this method for measuring CBF in patients with severe head injury. The technique is inexpensive and can be done at the bedside a minimal amount of specialized equipment. The diffusible indicator, nitrous oxide, is widely available and has little effect on cerebral hemodynamics in the low concentrations used. The measurements can be repeated as often as is necessary. In addition, since the venous blood samples for measurements of arteriovenous differences of oxygen, glucose, and lactate are obtained from the same distribution in the brain as the CBF measurement, the calculations of cerebral metabolism (CMR = AVD × CBF) are, at least theoretically, more accurate. The major disadvantage of the nitrous oxide method is that no information is obtained about regional variations in CBF.

Results

CBF and AVDO$_2$ after Head Injury

CBF was normal in 47 patients, averaging 0.41 ± 0.10 ml/gm/min. Twenty-five patients had a reduced CBF with a mean value of 0.29 ± 0.05 ml/gm/min, and 30 had an elevated CBF, averaging 0.62 ± 0.14 ml/gm/min. The differences in CBF were reflected by differences in AVDO$_2$, which averaged 2.1 ± 0.7 µmol/ml in the group with a reduced CBF, 1.9 ± 0.5 µmol/ml in those with a normal CBF, and 1.6 ± 0.5 µmol/ml in patients with an elevated CBF. The differences not only in CBF level but also in the ratio between CBF and cerebral metabolism, as reflected by the AVDO$_2$, suggested separation of the patients into groups with distinct pathophysiologies.

Relationship of CBF and Neurological Outcome

In the patients with a reduced CBF, the mortality was 32%, compared to 21% in patients with a normal CBF and 20% in patients with an elevated CBF. The patients with a reduced CBF were more likely to have an outcome of severe disability or persistent vegetative state, 48% compared to 39% of patients with a normal CBF and 27% of patients with an elevated CBF. Twenty percent of patients with a reduced CBF had a good recovery or moderate disability compared to 53% of patients with an elevated CBF and 41% of patients with a normal CBF.

To determine whether the differences in neurological outcome among the 3 groups reflected the effect of other covariates, logistic regression analysis was used. The model was fitted in a forward selection manner, with CBF group, age, and GCS forced into the model in the first step. The final model (Table 1) contained only those additional variables that were significantly associated with outcome. When the effect of CBF on neurological outcome was adjusted for age, initial GCS, hemoglobin concentration, cerebral perfusion pressure, and CMRO$_2$, a reduced

Table 1. *Logistic Regression Model of the Relationship of Cerebral Blood Flow to Neurological Outcome, with Dependent Variable of Favorable Outcome at 3 Months after Injury*

Predictor	Coefficient	Chi-Square	p Value
CBF Group			0.04
elevated	1.00	6.15	
normal	−0.12	0.13	
reduced	−0.88	3.94	
Age	−0.03	1.81	0.18
Initial GCS			0.02
3–6	−0.60	5.86	
>6	0.60	5.86	
Cerebral oxygen consumption	3.44	6.87	0.01
Cerebral perfusion pressure	−0.04	3.68	0.06
Hemoglobin concentration	−0.65	6.38	0.01
Intercept	8.32	5.40	0.02

CBF was still significantly associated with an unfavorable neurological outcome.

Discussion

The present study suggests that in addition to the factors of age, neurological status, and raised ICP, the level of CBF may be a factor in determining neurological outcome. The cause of the poorer outcome in the patients with reduced CBF is not clear. The association could not be explained on the basis of any identifiable confounding factor. The mean CBF for the group with reduced CBF was 0.29 ± 0.05 ml/gm/min, which is well above ischemic thresholds for the brain. These ischemic thresholds were determined in normal experimental animals, and it may not be appropriate to extrapolate these data directly to patients with a neurological injury. It might be hypothesized that a low CBF does not allow for adequate compensation during periods of stress. Transient periods of hypoxia, hypotension, or intracranial hypertension might be more likely to result in ischemia in patients with reduced CBF reserve, adversely affecting outcome.

References

1. Fieschi C, Battistini N, Beduschi A, *et al* (1974) Regional cerebral blood flow and intraventricular pressure in acute head injuries. J Neurol Neurosurg Psychiatry 37: 1378–1388
2. Kety SS, Schmidt CF (1948) The nitrous oxide method for the quantitative determination of cerebral blood flow in man: theory, procedure, and normal values. J Clin Invest 27: 476–483
3. Muizelaar JP, Marmarou A, DeSalles AAF, *et al* (1989) Cerebral blood flow and metabolism in severely head-injured

children. Part 1. Relationship with GCS score, outcome, ICP, and PVI. J Neurosurg 71: 63–71

4. Overgaard J, Molsdal C, Tweed WA (1981) Cerebral circulation after head injury. Part 3. Does reduced regional cerebral blood flow determine recovery of brain function after blunt head injury? J Neurosurg 55: 63–74

5. Overgaard J, Tweed WA (1983) Cerebral circulation after head injury. Part 4. Functional anatomy and boundary-zone flow deprivation in the first week of traumatic coma. J Neurosurg 59: 439–446

6. Robertson CS, Contant CF, Gokaslan ZL, Narayan RK, Grossman RG (1992) Cerebral blood flow, arteriovenous oxygen difference, and outcome in head injured patients. J Neurol Neurosurg Psychiatry 55: 594–603

7. Tabaddor K, Bhushan C, Pevsher PH, *et al* (1972) Prognostic value of cerebral blood flow (CBF) and cerebral metabolic rate of oxygen (CMRO$_2$) in acute head trauma. J Trauma 12: 1053–1055

Correspondence and Reprints: Claudia Robertson, M.D., Department of Neurosurgery, Baylor College of Medicine, One Baylor Plaza, Houston, TX 77030, U.S.A.

Acta Neurochir (1993) [Suppl] 59: 28–33

Xenon 133 – CBF Measurements in Severe Head Injury and Subarachnoid Haemorrhage

J. Meixensberger

Department of Neurosurgery, University of Würzburg, Würzburg, Federal Republic of Germany

Summary

The possibility of measuring cerebral blood flow by mobile bedside units with the intravenous 133-Xenon technique increased the interest to monitor haemodynamic changes after head injury and subarachnoid haemorrhage in intensive care.

Time course of resting CBF after trauma is variable (reduced CBF, hyperemia) and there is no strong correlation to clinical outcome. Additional studies of CBF/CO_2 reactivity show normal and impaired CO_2 response in the acute stage after trauma (day 1–8). A permanently impaired CO_2 reactivity correlates with severe brain damage and bad outcome (GOS 1,2). A normal or improving CO_2 reactivity indicates a favourable outcome (GOS 3–5).

There was no significant correlation between CBF and ICP, nor between CBF and CPP. A CPP of more than 70 mmHg did not guarantee a sufficient CBF in every case indicating the variability of the limits of autoregulation. As therapeutic hyperventilation may lead to ischemia, mannitol was preferred to reduce ICP and increased low CBF to normal values. This fact should be considered in the treatment of patients with low CBF and normal CO_2 reactivity.

Delayed ischemic neurological deficits ("vasospasm") are well – known as significant complications of the clinical course following SAH. Immediately postoperatively performed CBF measurements enable to detect ischemia and allow to start early antiischemic therapy. During "vasospasm" CBF showed a better correlation to the neurological status than blood flow velocity in the basal arteries measured by transcranial doppler sonography. Futhermore hyperemia after SAH could only be verified by CBF measurements.

Keywords: 133 Xenon cerebral blood flow; transcranial doppler ultrasonography; head injury; subarachnoid haemorrhage.

Introduction

In the last years progress in neurointensive care has improved the prognosis of head injury and subarachnoid haemorrhage (SAH)[11, 19]. Vascular events like secondary hyperemia or ischemia in the acute phase of illness are important for the outcome of such patients. The possibility of measuring cerebral blood flow (CBF) by mobile bedside units increased the interest to monitor such haemodynamic changes and to evaluate the usefulness in predicting clinical outcome and benefit of different therapeutic principles. The major purpose of CBF studies in the acute phase of traumatic brain injury or SAH is to recognize insufficient CBF and to start an adequate therapy immediately. The following issues on haemodynamic changes after head injury are thus addressed:

- time course of CBF and CBF/CO_2 reactivity and prognostic value
- relationship between CBF and intracranial pressure (ICP) as well as cerebral perfusion pressure (CPP)
- risk to cause ischemia by various treatment strategies

Furthermore, secondary haemodynamic events after SAH will be discussed answering the following questions:

1. Can CBF measurements detect early complications after operation?
2. Is there a correlation between transcranial doppler sonography (TCD) blood flow velocity and CBF in the time course of vasospasm?

Patients and Methods

A total of 278 CBF measurements was performed in 80 patients in the intensive care unit (42 patients after head injury, 38 patients after SAH) during day 1 to 14 after trauma, or day 0 to 22 after SAH respectively. Patient's age ranged from 7–79 years, with an average of 40.6 years. On admission, head injured patients had a GCS score less than 8, SAH patients were classified to the Hunt and Hess grading scale[10] and were mostly

HH 1 to 3. Head injured patients were always sedated, intubated and ventilated. Immediate postoperative studies after aneurysm clipping were done under sedation, intubation and ventilation, later with spontanous breathing.

A mobile Cerebrograph (Novo 10a, Copenhagen, Denmark) with 10 detectors, five over each hemisphere, was utilized for the CBF measurements. After intravenous administration of 10–30 mCi of Xenon 133 regional CBF was measured. The initial slope index (ISI)[25] or the CBF 15 by the Obrist model[23] were determined. Whenever possible two CBF measurements were performed at different $PaCO_2$ levels in order to determine rest CBF under the actual therapeutic conditions and CBF-reactivity to $PaCO_2$ changes. Global and hemispheric CBF as well as global and hemispheric CBF/CO_2 reactivity (CBF change per mmHg $PaCO_2$ change were calculated). A range from 35 to 59 ISI units or a CBF 15 from 32.9 to 55.2 ml/100 g/min was defined a normal rest CBF[18, 22]. According to Maximilian a change of more than 0.75 ISI units per mmHg change was regarded as normal[17]. With each CBF measurement, the arterial pCO_2 was checked. Whenever possible transcranial doppler sonography was used (TCD 2000 EME Überlingen, Federal Republic of Germany) to measure blood flow velocities in the major basal arteries (ICA, MCA). Additionally, intracranial pressure (Gaeltec transducer), arterial blood pressure, and body temperature were continuously monitored. The neurological status was documentated daily. Outcome data of head injured patients were evaluated 12 months after injury.

Results

CBF after Head Injury: Rest CBF and CBF/CO_2 Reactivity

Following head injury, global and hemispheric CBF values were reduced or subnormal in about 55% of the cases, in 45% of the measurements hyperaemic. Sometimes CBF normalized in the following course. Correlating the CBF data to outcome 12 months after trauma three groups could be detected (Fig. 1): There are patients with permanently reduced CBF and bad outcome (GOS 1,2) and patients with normal CBF and favourable outcome (GOS 3–5). A third group had a variable outcome and variable global rest CBF.

The analysis of the CBF/CO_2 reactivity (n = 68) as an additional indicator for CBF dysregulation revealed a reduced or inverse CBF/CO_2 reactivity in 41% of the cases during the first eight days after trauma. Correlating these data again to GOS 12 months after injury the following subgroups could be distinguished (Fig. 2): Patients with a permanently reduced CBF/CO_2 reactivity had a bad outcome (GOS 1,2) and patients with a normal reactivity reached GOS 3 to 5. However, there were patients with favourable outcome (GOS 3–5) who improved from subnormal to normal cerebrovascular reactivity.

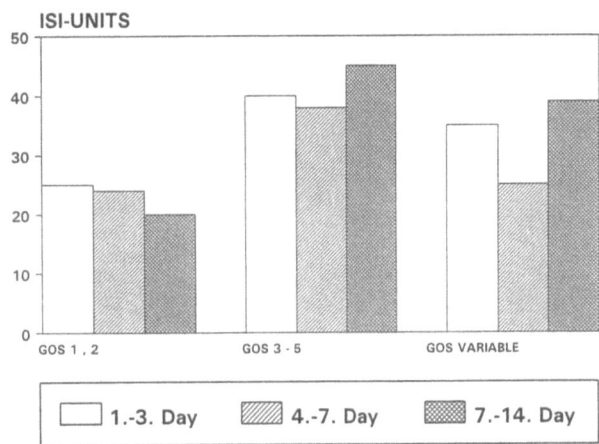

Fig. 1. Global CBF (ISI units) at rest during the acute stage of brain injury (day 1–14) and clinical outcome (GOS) 12 months after trauma in head injured patients (n = 35). Values given as mean values. Normal CBF range from 35 to 59 ISI units

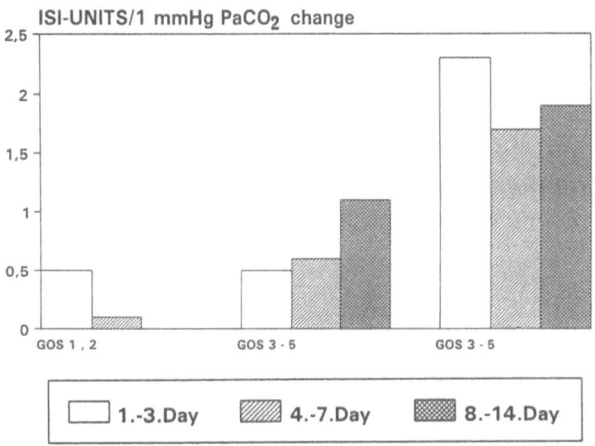

Fig. 2. CBF/CO_2 reactivity (ISI units/mmHg $PaCO_2$ change) during the acute stage of brain injury (day 1–14) and clinical outcome (GOS) 12 months after trauma in head injured patients (n = 32). Values given as mean values. Normal CBF/CO_2 reactivity >0.75 ISI units/l mmHg $PaCO_2$ change

Correlation between Global CBF and ICP or CPP

There was no significant difference between the ICP in the hyperemia group and the group with reduced cerebral blood flow (hyperemia group: ICP 21 ± 11 mmHg, group with reduced flow: ICP 23 ± 14 mmHg). Furthermore, no significant correlation between CBF and CPP was found (Fig. 3). A cerebral perfusion pressure of above 70 mmHg did not guarantee a normal CBF in all cases.

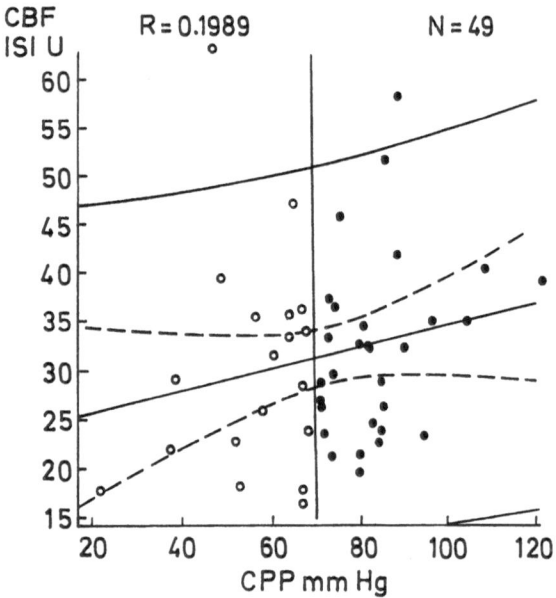

Fig. 3. Lack of correlation between global CBF (ISI units) and cerebral perfusion pressure (CPP) (mmHg) after head injury. Simultaneous measurements n = 49

Haemodynamic Effects of Hyperventilation and Mannitol

If the CBF/CO_2 reactivity was intact, there was a risk to critically decrease CBF by hyperventilation therapy. The group with reduced cerebral blood flow at rest was particularly susceptible to develop ischemia during hyperventilation. Figure 4 summarizes the effects of hyperventilation in the group with reduced cerebral blood flow and the hyperemia group. On the other hand, mannitol increased a reduced CBF as shown in Fig. 5.

CBF after Subarachnoid Haemorrhage and after Aneurysm Surgery

In four of thirty two SAH patients ischemic CBF patterns were found after aneurysm surgery. Table 1 summarizes the data of the immediately postoperatively performed CBF measurements. In patients with ischemia a statistically significant interhemispheric side difference (>10%) with a decreased CBF on the affected side was detected. In addition, there was a significantly lower and mostly subnormal CBF/CO_2 reactivity in ischemia patients indicating a reduced cerebrovascular reserve capacity.

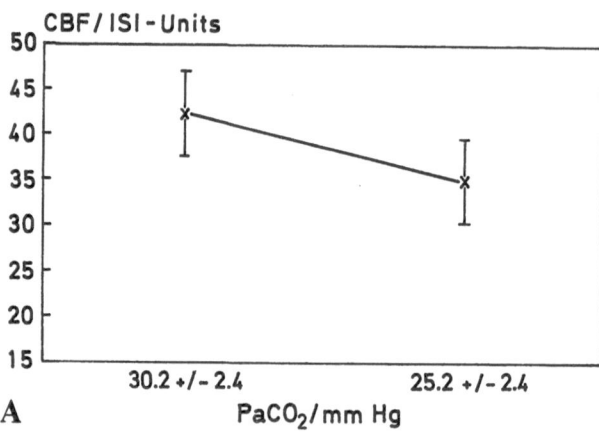

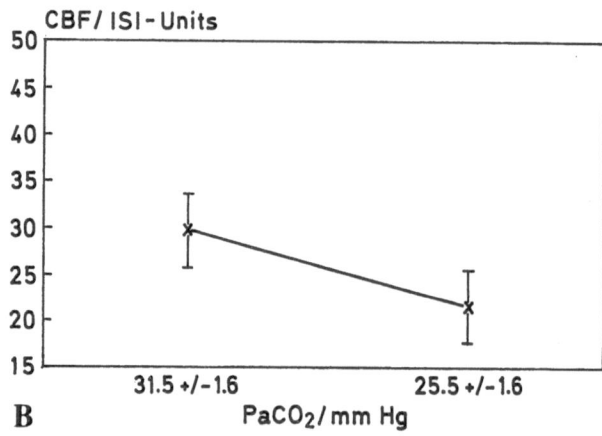

Fig. 4. Effect of hyperventilation on global CBF in the hyperemia group (A) and the group with reduced cerebral blood flow (B) with normal CBF/CO_2 reactivity. Values are given as mean ± SD. Hyperemia group n = 20, group with reduced cerebral blood flow n = 15

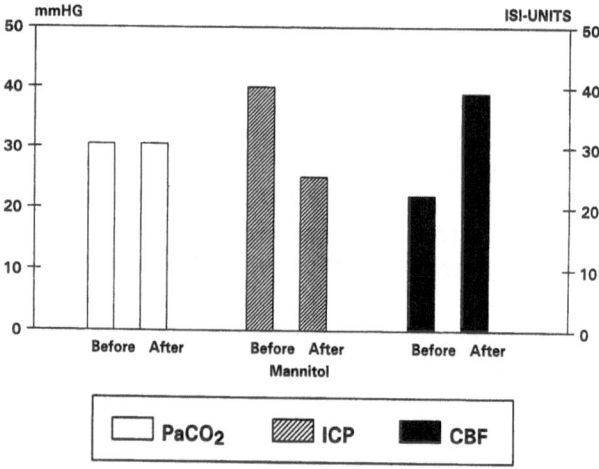

Fig. 5. Effect of mannitol (1 g/kg body weight i.v.) on global CBF (ISI units) and mean ICP (mmHg) in a head injured patient (male, 40 years, diffuse brain edema, 4th day after injury)

Table 1. *CBF at Rest (ISI Units), Interhemispheric Side Difference (%) and CBF/CO$_2$ Reactivity (ISI Units/mm Hg PaCO$_2$ Change) Immediately after Aneursym Surgery*

	Rest – CBF OP	no OP	% Side difference	CBF CO$_2$ reactivity OP	no OP
Ischemia n = 4	28.1 ± 6.3	30.7 ± 9.0	11.5 ± 2.5	0.5 ± 0.6*	0.8 ± 0.5*
No ischemia n = 28	40.1 ± 12.8	40 ± 12.8	3 ± 1.4	1.9 ± 1.2	1.7 ± 1.1

Patients with (n = 4) and without (n = 28) ischaemia postoperatively. Op. = side of operation, no Op. = contralateral side. * p < 0.01.

Critical TCD Flow velocity 180cm/sec MCA

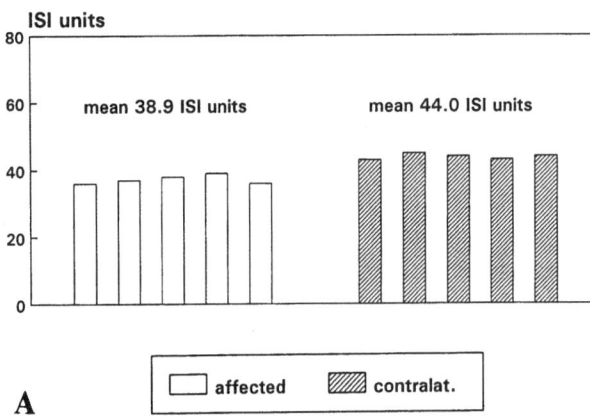

A

Critical TCD Flow velocity 160cm/sec MCA

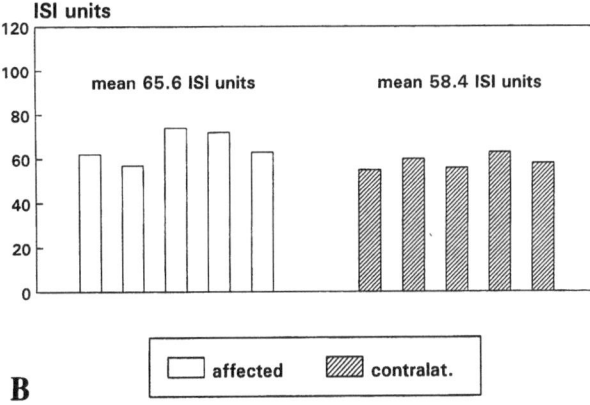

B

Fig. 6. TCD flow velocity in middle cerebral artery (cm/sec) after SAH. (A) Vasospasm; (B) hyperaemia detected by simultaneous CBF measurements (ISI units). Affected: side of operation and increased blood flow velocity; contralat.: contralateral side. CBF values are given as hemispheric mean ± SD

Correlation of CBF and TCD in the Course of Vasospasm

There was no significant correlation between the absolute ISI values of the MCA region and respective blood flow velocity values (n = 62, r = 0.29). In general, higher flow velocities were seen on the side of the ruptured aneurysm than on the contralateral side. A critical blood flow velocity above >120 cm/s was not always related to vasospasm. Also hyperaemia was seen on the side of a critically increased blood flow velocity (Fig. 6). Regarding the interhemispheric difference there was a correlation between CBF and blood flow velocity in 75% of our measurements (r = 0.78, p < 0.001). On the other hand there were patients with significant interhemispheric differences for TCD measured velocities in the MCA without any significant interhemispheric CBF difference and vice versa. Figure 7 shows CBF and TCD values of six patients with delayed neurological deficits in the course after SAH. At the maximum of the delayed neurological deficit the blood flow velocity in the MCA on the side of the ruptured aneurysm was always critically increased and a corresponding CBF reduction was found. During clinical improvement the CBF values normalized in all cases, but critical blood flow velocities were seen in three of six cases. While there was a good correlation between clinical vasospasm and CBF, there was a lesser correlation between vasospasm and TCD data.

Discussion

Mobile, bedside CBF units with multiple detectors using the intravenous 133 Xenon technique

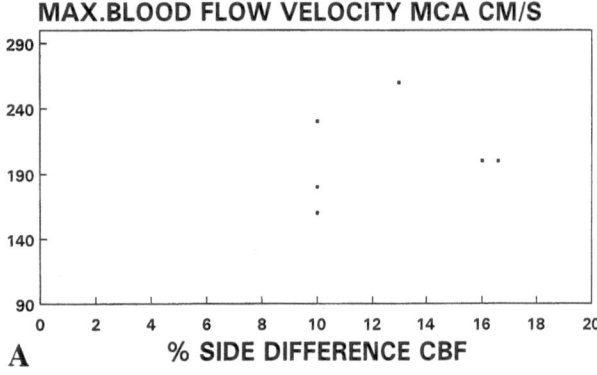

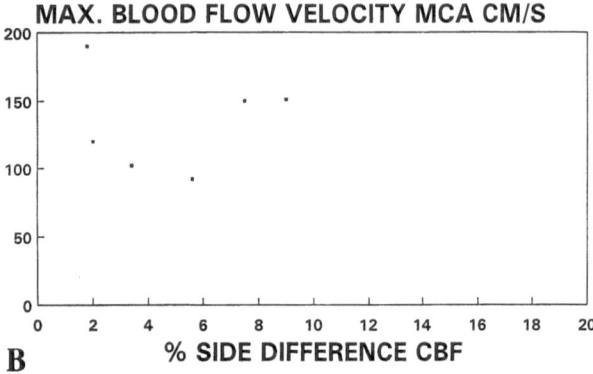

Fig. 7. Side difference CBF (significant >10%) and TCD (blood flow velocity in the middle cerebral artery (MCA) cm/sec) in the time course after SAH. (A) Patients with delayed neurological deficits (5th to 14th day after SAH). (B) Same patients without neurological deficits (9th to 22nd day after SAH)

allow CBF measurements in neurosurgical intensive care. In addition, this method makes repeated measurements possible to study hemodynamic changes and influences of therapeutic strategies.

In the course after trauma global and hemispheric CBF and CBF/CO_2 reactivity varies depending on the severity of tissue damage, intracranial hypertension etc.[3-6, 21, 22, 24]. The CBF data can, however, not predict the outcome of injured patients[16, 21, 22]. Uncoupling of CBF and metabolism might be one reason. Lassen described this as "luxury perfusion syndrome" in patients with acute brain disorders[13]. CBF/CO_2 reactivity seems to be a better prognostic parameter[6-8, 18, 24]. Permanent impaired CO_2 response correlates with severe brain damage and poor outcome. Improvement of impaired CO_2 reactivity is possible and often correlates with clinical improvement.

Confirming results of others there is no correlation between ICP and CBF. Obviously ICP and CBF are not direct related[3, 5, 14, 20, 24]. Since maintenance of cerebral perfusion pressure above

70 mmHg does not guarantee a sufficient CBF in every case continuous monitoring of CBF (e.g. laser doppler flow, jugular venous oxygen saturation) is necessary[15]. In addition, more should be known about the hemodynamic effects of various treatment regimens. CBF measurements allow to recognize patients who are at risk for ischaemia, e.g. by hyperventilation. Treatment of increased intracranial pressure by mannitol can increase CBF and diminish the risk of secondary ischaemia in those patients. This might improve the clinical course and the outcome of such patients.

Complications (ischaemia, vasospasm) after aneurysm surgery known to deteriorate the prognosis after SAH[10, 11] can be detected by CBF measurements and can be counteracted.

Transcranial Doppler ultrasonography is an additional noninvasive method to assess flow velocity in basal arteries[1]. Though CBF and blood flow velocity is well correlated under normal physiological conditions[2], there is not always a sufficient correlation in vasospasm. Critically increased blood flow velocities in the basal arteries indicating vasospasm[9, 12, 26, 27] can also be caused by hyperemia. CBF measurements are sensitive to detect such changes.

Cerebral blood flow and additional metabolic monitoring of the damaged brain are important to enhance our knowledge of various cerebral pathologies. Perhaps new technics, like monitoring of CBF by laser doppler probes and of cerebral metabolism may increase our knowledge and improve our treatment regimens. 133-Xenon CBF studies are, however, still necessary to validate such new methods of monitoring cerebral blood flow.

References

1. Aaslid R, Markwalder TM, Nornes H (1982) Noninvasive transcranial Doppler ultrasound recording of flow velocity in basal cerebral arteries. J Neurosurg 57: 769–774
2. Brass LM, Prohovnik I, Pavlakis S, Mohr JP (1987) Transcranial Doppler examination of middle cerebral artery velocity versus xenon rCBF: two measures of cerebral blood flow. Neurology 37 [Suppl 1]: 85
3. Bruce DA, Langfitt TW, Miller JD, *et al* (1973) Regional cerebral blood flow, intracranial pressure and brain metabolism in comatose patients. J Neurosurg 38: 131–144
4. Cold GE, Jensen FT, Malmros R (1977) The effects of PaCO2 reduction on regional cerebral blood flow in the acute phase of brain injury. Acta Anaesthesiol Scand 21: 359–367
5. Enevoldsen EM, Cold G, Jensen FT, *et al* (1976) Dynamic changes in regional CBF, intraventricular pressure. CSF pH and lactate levels during the acute phase of head injury. J Neurosurg 44: 191–214

6. Enevoldsen EM, Jensen FT (1978) Autoregulation and CO_2 responses of cerebral blood flow in patients with acute severe head injury. J Neurosurg 48: 689–703

7. Fieschi C, Battistini N, Beduschi A, et al (1974) Regional cerebral blood flow and intraventricular pressure in acute head injuries. J Neurol Neurosurg Psychiatry 37: 1378–1388

8. Gordon E, Bergvall U (1973) The effect of controlled hyperventilation on cerebral blood flow and oxygen uptake in patients with brain lesions. Acta Anaesthesiol Scand 17: 63–69

9. Harders AG, Gilsbach JM (1987) Time course of blood velocity changes related to vasospasm in the circle of Willis measured by transcranial Doppler ultrasound. J Neurosurg 66: 718–728

10. Hunt WE, Hess RM (1968) Surgical risk as related to time of intervention in the repair of intracranial aneurysms. J Neurosurg 28: 14–20

11. Kassell NF, Saski T, Colohan ART, Nazar G (1985) Cerebral vasospasm following aneurysmal subarachnoid haemorrhage. Stroke 16: 562–72

12. Klingelhöfer J, Sander D, Holzgraefe M, et al (1991) Cerebral vasospasm evaluated by transcranial Doppler ultrasonography at different intracranial pressures. J Neurosurg 75: 752–758

13. Lassen NA (1966) The luxury perfusion syndrome and its possible relation to acute metabolic acidosis localized within the brain. Lancet 2: 1113–1115

14. Langfitt TW (1976) Incidence and importance of intracranial hypertension in head injured patients. In: Beks JWF, Bosch DA, Brock M (eds) Intracranial Pressure III. Springer, Berlin Heidelberg New York, pp 67–72

15. Langfitt TW, Weinstein JD, Kassell NF (1965) Cerebral vasomotor paralysis produced by intracranial hypertension. Neurology 15: 622–641

16. Langfitt TW, Obrist WD, Gennarelli TA, et al (1977) Correlation of cerebral blood flow with outcome in head injured patients. Ann Surg 186: 411–414

17. Maximilian VA, Prohovnik I, Risberg J (1980) The cerebral hemodynamic response to mental activation in normo- and hypercapnia. Stroke 11: 342–347

18. Messeter K, Nordström CH, Sundberg G, et al (1986) Cerebral hemodynamics in patients with acute severe head trauma. J Neurosurg 64: 231–237

19. Miller JD (1985) Head injury and brain ischemia – implications for therapy. Br J Anaesth 57: 120–129

20. Muizelaar JP, Marmarou A, DeSalles AAF, Ward JD, Zimmerman RS, Li Z, Choi SC, Young HF (1989a) Cerebral blood flow and metabolism in severely head-injured children. Part 1: Relationship with GCS score, outcome, ICP, and PVI. J Neurosurg 71: 63–71

21. Obrist WD, Gennarelli TA, Segawa H, et al (1979) Relation of cerebral blood flow to neurological status and outcome in head injured patients. J Neurosurg 51: 292–300

22. Obrist WD, Langfitt TW, Jaggi JL, et al (1984) Cerebral blood flow and metabolism in comatose patients with acute head injury. J Neurosurg 61: 241–253

23. Obrist WD, Thomson HK Jr, Wang HS, et al (1975) Regional cerebral blood flow estimated by [133]Xenon inhalation. Stroke 6: 245–256

24. Overgaard J, Tweed WA (1974) Cerebral circulation after head injury. Part I. Cerebral blood journal and its regulation after closed head injury with emphasis on clinical correlation. J Neurosurg 41: 531–541

25. Risberg J, Ali Z, Wilson EM, et al (1975) Regional cerebral blood flow by [133]Xenon inhalation. Preliminary evaluation of an initial slope index in patients with unstable flow compartments. Stroke 6: 142–148

26. Seiler RW, Grolimund P, Aaslid R, et al (1986) Cerebral vasospasm evaluated by transcranial ultrasound correlated with clinical grade and CT-visualized subarachnoid hemorrhage. J Neurosurg 64: 594–600

27. Sekhar LN, Wechsler LR, Yonas H, et al (1988) Value of transcranial Doppler examination in the diagnosis of cerebral vasospasm after subarachnoid hemorrhage. Neurosurgery 22: 813–821

Correspondence and Reprints: Dr. J. Meixensberger, Neurochirurgische Klinik und Poliklinik der Universität Würzburg, Josef-Schneider-Strasse 11, D-97080 Würzburg, Federal Republic of Germany.

Acta Neurochir (1993) [Suppl] 59: 34–40

Evaluation of Regional Cerebral Blood Flow in Acute Head Injury by Stable Xenon-Enhanced Computerized Tomography

G. J. Bouma[1] and **J. P. Muizelaar**[2]

[1] Department of Neurosurgery, University of Amsterdam, The Netherlands, and [2] Division of Neurosurgery, Medical College of Virginia, Richmond, Virginia, U.S.A.

Summary

Measurement of regional cerebral blood flow (rCBF) in head-injured patients is considered useful for understanding the cerebral hemodynamics of brain trauma and for determining the optimal therapy. Most data thus far obtained with ^{133}Xe clearance techniques have made only relative contribution, due to limitations of the ^{133}Xe method. More recently, is has become possible to measure rCBF by xenon-enhanced computerized tomography (Xe-CT), which obviates most problems inherent to the ^{133}Xe method. On the other hand, computational errors and concerns regarding the safety of xenon inhalation have thwarted the clinical use of Xe-CT. Recent advances in CT technology, however, have largely eliminated these problems.

Xe-CT CBF measurements in severe head injury demonstrate a good correlation between CBF values obtained with ^{133}Xe and Xe-CT. By consistently applying these studies in conjunction with conventional CT, information on very early flow derangements (within 1 to 2 hours after injury) can be obtained, in relation to anatomical lesions. Preliminary data reveal higher incidences of global and focal ischaemia than found previously. Local ischaemia tends to evolve to hyperemia in the ensuing days.

Keywords: Head injury; hyperventilation; ischaemia; regional cerebral blood flow; xenon-enhanced computerized tomography.

Introduction

Disturbances of intracranial pressure (ICP) and cerebral blood flow (CBF) are common after severe head injury and are thought to be closely related to the clinical course and outcome. A correlation between the level of ICP and outcome has been clearly established[23, 27], and a clinical management strategy based on ICP monitoring and treatment of raised ICP has generally been recommended[3]. Similarly, it has been suggested that frequent determination of CBF may be of value in the treatment of severely head-injured patients[17] and accordingly, many reports on posttraumatic CBF measurement have been published, in most of which ^{133}Xe clearance techniques were utilized. In general, the results of these studies have been somewhat disappointing in terms of clinical impact[4]. A wide variation of CBF values has been found with no apparent correlation with clinical status or outcome[6, 9, 11, 28, 29, 32]. More recently, however, it was shown that a correlation between the level of CBF and clinical status and outcome does exist, but only during the first few hours following injury[4]. It has also been recognized that limitations of the ^{133}Xe method, such as its poor spatial resolution and "look-through" artifacts[31], are the reason that regional abnormalities in cerebral perfusion cannot be properly assessed[26, 28, 38]. Therefore, the clinical value of CBF measurement in head injury has been questioned, and it was postulated that valid observations on CBF in trauma had to await the development of new methods able to measure regional flow with high spatial resolution, preferably in a three-dimensional or tomographic fashion[9, 26]. Sophisticated techniques such as positron emission tomography (PET) or single-photon emission computed tomography (SPECT) may meet such requirements, but are not widely available and considered too impractical for routine use in critically ill patients. Consequently, very few studies on the use of PET and SPECT in head injury have been reported[1, 2, 37].

More recently, it has become possible to measure CBF by stable xenon-enhanced computerized tomography (Xe-CT)[13]. This method yields high-resolution, quantitative tomographic regional CBF (rCBF) data which is readily available and corresponds with the anatomical CT image. The ability to record CBF during CT scanning has important logistic advantages for the application in severely head-injured patients, because unnecessary transportation of the patient to other facilities is avoided.

In this paper, we will discuss some methodological aspects of the Xe-CT technique with regard to its use in head injury. Subsequently, we will review our experience with the application of Xe-CT in head-injured patients.

Methodology

Measurement of CBF by xenon-enhanced dynamic CT scanning was developed in the early 1980's[8, 13, 25]. It involves repeated CT scanning during the inhalation of a gas mixture containing stable (non-radioactive) xenon and oxygen. The inert xenon gas diffuses into the brain tissue immediately, the rate of tissue uptake being proportional to flow. Because xenon appears radiodense on CT, its concentration in the tissue can be monitored pixel-wise, and flow can be calculated from mathematical analysis of the uptake curve. By translating the flow value for each pixel into the equivalent Hounsfield unit, a quantitative flow map is created with high spatial resolution, corresponding with the anatomical CT image. Blood flow can be obtained for any region of interest (ROI), which can be chosen at will from the flow map. An in-depth discussion of the theoretical aspects of the Xe-CT method and the mathematical approach to the analysis of the xenon uptake curves can be found in special reports[10, 19, 34].

The Xe-CT technique has several distinct advantages over other methods for CBF measurement, especially with regard to head injury. First, its high spatial resolution along with the ability to record zero flow[40] allows for the identification of relatively small areas of ischaemia. Second, blood flow in both superficial cortex and deeper structures, such as the brainstem and the basal ganglia, can easily be assessed, while the technique provides direct correlation of flow to anatomy as visualized on CT. This permits easy recognition of the tissue being investigated, as well as quantification of flow in morphologically abnormal areas (e.g. contusions or edema). In addition, such regions can be easily re-identified in subsequent studies, to evalute temporal changes in regional flow in relation to pathophysiological processes. Third, the method is repeatable, so that CBF responses to various haemodynamic challenges can be assessed, which may provide important information on vascular reactivity, autoregulation, and cerebrovascular reserve[41]. Finally, Xe-CT is easily and rapidly performed in conjunction with conventional CT, requiring relatively simple additional equipment and software only, which makes the technique well-suited for assessment of CBF in the acute stage of neurologic emergencies such as trauma and stroke[5, 16].

On the other hand, several limitations of the technique should be considered. Computational errors may occur due to CT noise, tissue heterogeneity, motion artifacts and inaccuracies in the curve-fitting procedure, which may result in deviations exceeding 100% of estimated flow in a single pixel[12, 14]. However, such errors can be overcome by choosing ROIs of at least $100\,mm^2$ and by excluding areas of high error as indicated on the error map[14]. Head motion usually can be avoided in severely head-injured patients by muscle paralysis, artificial ventilation, and proper fixation of the head. It should also be noted that CBF is measured in only two or three levels, which limits each study to a brain volume of only 20–25% of the total brain. Thus, even if normal flow is found in all areas, flow abnormalities in regions that are not studied cannot be excluded.

The most important issue appertains the possible adverse effects of xenon-inhalation. With Xe-CT, high tissue concentrations of xenon are needed to achieve a sufficient signal-to-noise ratio, and therefore, xenon needs to be administered in much higher doses than with the ^{133}Xe method. Concerns have been raised that xenon inhalation in these high concentrations for prolonged periods of time might not be tolerated. In concentrations greater than 50%, xenon is a general anesthetic, and may cause respiratory depression[39]. In addition, xenon causes cerebral vasodilation which may lead to undesired side-effects such as increases in CBF and ICP, which may be further aggravated by increased airway pressure[7]. Anecdotal reports of such problems seemed to corroborate these concerns and have slowed the acceptance of the technique in clinical practice[14, 15, 39]. However, recent studies have demonstrated that the inhalation of 32% stable xenon for 4.5 minutes is safe[19] and does not increase ICP or CBF significantly[7, 21, 36]. Nevertheless, these concerns warrant close observation of the patient during the examination which includes monitoring of blood pressure, end-tidal CO_2, oxygen saturation, and preferably ICP. Thus far, only a small number of studies of Xe-CT CBF measurements in head injury has been reported[5, 7, 15, 18, 20, 22].

Results

Correlation of Xe-CT CBF with ^{133}Xe CBF in Head-Injured Patients

The Xe-CT method was compared with the non-invasive ^{133}Xe technique in 13 severely head-injured patients. A total of 16 ^{133}Xe-CBF (i.v. injection method) and stable Xe-CT CBF studies were obtained within 1 to 4 hours (mean: 3.1 hours) from each other. The Xe-CT studies were performed on a GE-9800 CT scanner using standard hardware and software supplied by the manufacturer (General Electric, Milwaukee, WI). The patients inhaled 32% stable xenon in oxygen for 4.5 minutes, administered through a volume-cycled respirator. The ^{133}Xe CBF measurements were performed using a portable system (Novo-10a Cerebrography, Novo Diagnostics, Bagsvaerd, Denmark) containing 2 fixed sets of 5 detectors, which are positioned over each hemisphere covering the frontal, temporal and parietal areas. From the washout curves CBF_{15} was calculated, a non-compartmental index of flow, which has shown greater stability with severe brain injury than traditional two-compartment models[30]. For the comparison between the Xe-CT and ^{133}Xe methods, the mean cortical flow in the regions underlying the gamma detectors, was calculated for the corresponding areas on the flow maps obtained with Xe-CT and correlated against CBF_{15}. $PaCO_2$-differences between studies were corrected by assuming a 3% change in CBF per torr CO_2.

We found a good correlation between CBF values

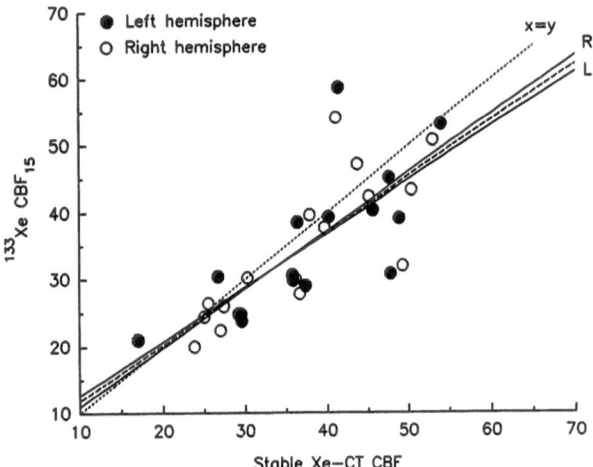

Fig. 1. Graph showing the comparison between CBF values obtained with ^{133}Xe clearance and xenon-enhanced CT. The correlation coefficient r = 0.79, p < 0.001. The regression line did not differ significantly from the line of unity (x = y)

as measured by the two methods (Fig. 1), with an overall correlation coefficient r = 0.79 (p < 0.001); the regression line did not differ significantly from the line of unity (x = y). Thus, we found no evidence to support a significant flow-activation due to stable xenon inhalation. Moreover, the data indicate that the results of the Xe-CT studies can be evaluated in the light of the existing data on CBF in head injury acquired with [133]Xe.

Early Assessment of Regional CBF in Head-Injured Patients

In earlier studies of CBF in head-injured patients, a low incidence of ischaemic flow was found and no correlation between posttraumatic CBF levels and outcome was revealed[9, 29, 32]. Some investigators have therefore suggested that ischaemia, if present, only occurs during the first few hours after injury, and

has generally subsided when CBF is first measured[28]. This contention also concurs with a growing body of laboratory data indicating the importance of subcellular cascades leading to tissue damage that take place in the early phase after injury[24, 35]. Accordingly, in a previous study using ^{133}Xe, we had found ischaemia in one third of the patients in whom CBF was measured within 6 hours after injury[4].

To further pursue this matter, and specifically address the role of small areas of focal ischemia, in particular in the presence of intracranial mass lesions, we have made use of the above-mentioned advantages of the Xe-CT method. In 35 severely head-injured patients (Glasgow Coma Scale score ≤8) Xe-CT CBF measurements were performed at the time of the initial head CT scan obtained on admission to the hospital[5]. By this procedure, the time between injury and CBF measurement was minimized while it also enabled us, for the first time, to assess CBF in patients with intracranial mass lesions still in place. The average time after injury at which the studies were performed was 3.1 ± 2.1 hours. Flow values were obtained for both cerebral hemispheres (including gray and white matter), the brainstem, basal ganglia and the cerebellum. In addition, regional cortical flow was obtained for each of the cerebral lobes, using small circular ROIs of 2 cm diameter.

The results indicate that patients with diffuse brain swelling and patients with acute subdural haematomas had significantly lower flows and higher incidence of ischaemia in all measured brain regions than had patients with focal contusions, extradural haematomas or diffuse injuries without swelling (Table 1). Ischaemia was usually confined to supratentorial cortical areas, or was present as a global decrease in blood flow throughout the hemispheres. Isolated ischaemia in the brainstem or the basal

Table 1. *Results of Xenon-Enhanced CT CBF Measurements in 35 Severely Head-Injured Patients**

CT findings	No. of Cases	Global CBF	No. of Cases with Ischemia (%)
Acute SDH or brain swelling	15	25.8 ± 12.5[a]	9 (60)[b]
Other lesions	20	35.9 ± 7.3	2 (10)

* CT findings reflect the most prominent abnormality seen on the initial CT scan; *SDH* Subdural haematoma; other lesions include extradural haematomas, focal contusions and diffuse injuries without brain swelling; *CBF* cerebral blood flow (ml/100 g/min, mean ± standard deviation); Ischaemia is defined as CBF ≤ 18 ml/100 g/min.

The differences between the CT groups were statistically significant: [a] p < 0.005, one-tailed Student t-Test, [b] p < 0.005, Chi-square Test.

Table 2. *Regional Cerebral Blood Flow and Pupillary Reactivity in Head-Injured Patients**

Pupillary reactions	No. of Patients	Regional Cerebral Blood Flow		
		brainstem	basal ganglia	hemispheres
Normal	24	47.9 ± 12.2	42.8 ± 9.3	34.9 ± 8.1
Abnormal unilaterally	5	31.3 ± 15.9	40.0 ± 12.8	34.9 ± 8.4
Abnormal bilaterally	6	21.0 ± 12.7	22.1 ± 12.1	15.3 ± 8.6

* All values are mean ± standard deviation (ml/100 g/min); the differences in brainstem blood flow between patients with different patterns of pupillary reactivity were statistically significant (one-way analysis of variance, p < 0.002).

ganglia was not found. However, when pupillary asymmetry was present as a sign of tentorial herniation, brainstem blood flow was disproportionly reduced in relation to global CBF (Table 2). This might indicate impaired perfusion as a result of mechanical brainstem compression. Ischaemia was found in 11 out of 35 patients (31%), and was associated with early death (within 48 hours after injury). This could not be related to early hypoxic or hypotensive events, suggesting some other mechanism being responsible for early posttraumatic ischaemia. However, the sample sizes are still relatively small, and these data should therefore be considered preliminary.

Follow-up Studies

In the past, it has been difficult to establish a certain temporal pattern of the CBF changes occurring during the first days after traumatic brain injury. This was in great part due to the infrequent and often irregular timing of the CBF measurements, which usually cannot be performed more often for obvious medical and technical reasons. Thus, it is difficult to obtain closely-timed serial CBF studies that are necessary to resolve this issue, which is considered important for the understanding of the circulatory pathophysiology of severe traumatic brain injury. Ideally, this would require continuous CBF monitoring, for which, however, no satisfying technology yet exists.

Using Xe-CT, we were able to perform follow-up CBF studies at regular intervals during the first 4 days after injury in 24 patients. The time course of the average rCBF values in the measured areas for these patients is shown in Fig. 2. At the initial measurement, low values were found, but flow increased during the next days, reaching a peak on the second day post injury.

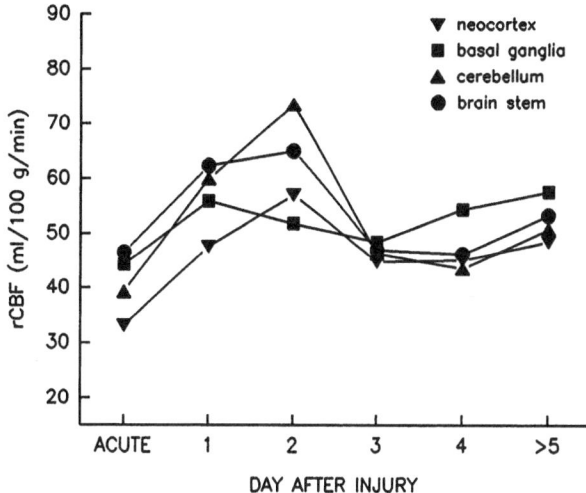

Fig. 2. Graph showing serial average rCBF values during the first 4 days after injury obtained in 24 head-injured patients

In general, flow in the various measured areas followed the same trend as the average global flow, although individual differences occurred. Notably, in regions with the lowest rCBF initially, the flow increase during the ensuing days was more pronounced. This is exemplified in Fig. 3, which shows the successive rCBF values of a patient with diffuse cerebral swelling and focal ischemic areas on the initial Xe-CT study. Eventually, this patient made a satisfying recovery. Figure 4 displays the initial and follow-up rCBF of a patient before and after removal of an extradural haematoma. Again, in the areas with the lowest flows initially strong flow increases were seen, progressing to frank hyperemia. Moreover, these hyperaemic areas exhibited a more pronounced response to hyperventilation, while in other areas this response was diminished or absent. Theoretically, such regional differences in cerebrovascular CO_2 reactivity might lead to unexpected adverse effects of hyperventilation, such as focal

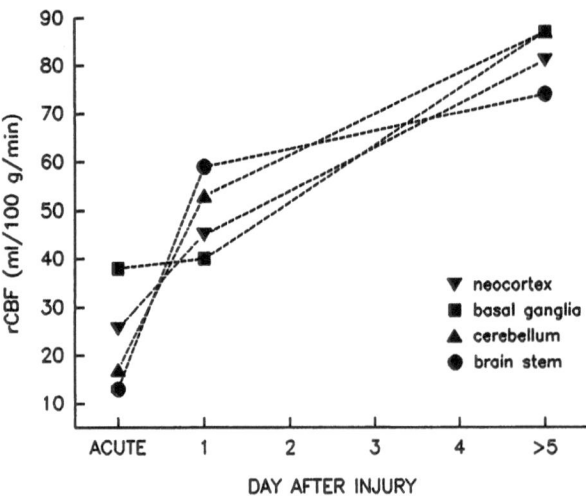

Fig. 3. Serial rCBF values in a patient with diffuse cerebral swelling. Focal ischaemia was reverted into hyperaemia during the first days after injury. The patient made a satisfying recovery (Glasgow Outcome Scale score 1 at 6 months post injury)

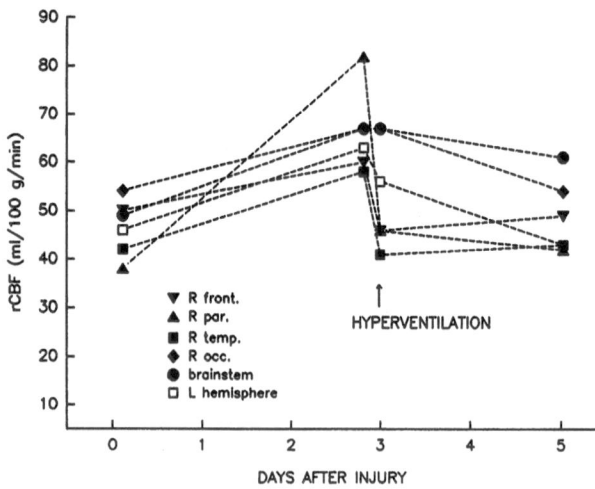

Fig. 4. Serial rCBF values in a patient before and after removal of an epidural haematoma. Hyperemia developed in the areas with the lowest flows initially, and stronger responses to hyperventilation were seen in these regions

ischaemia or worsening of local hyperemia due to "inverse steal"[33]. However, in an earlier study of CO_2 reactivity with double Xe-CT CBF measurements[20], we did not find such local paradoxical increases in flow with hypocapnia. Nonetheless, such studies may be helpful in determining the effect of hyperventilation and optimizing the therapeutic management in individual cases.

Discussion

The Xe-CT CBF method is a useful addition to the clinician's diagnostic instruments. It appears that these studies, when consistently applied in large series of head-injured patients with various types of intracranial pathology, may provide some answers to important questions pertaining to the cerebral circulatory pathophysiology of traumatic brain injury that could not be obtained previously. In particular, the fact that Xe-CT can be easily performed with any regular CT scan, gives the method an edge over more sophisticated techniques such as PET. Our preliminary results obtained in head injury appear promising in this regard.

However, certain aspects of the technique must be dealt with before the method can really be considered clinically useful. Most importantly, the potential harmful effects of xenon inhalation along with the fact that a complete study still takes about 15–20 minutes extra CT time (which may be too long for a patient in unstable condition) are matters of concern.

Another problem is the high cost of the xenon gas, which emphasizes the need for adequate clinical justification to perform these studies.

Hopefully, further advances in technology will help to overcome these problems. For instance, one could envision that faster and more sensitive CT scanners may allow for shorter inhalation times and further reduction of the necessary xenon concentrations[13]. Also, improvements in computer technology may help getting the results available more quickly or even in "real-time", so that clinical decision making can be truly based on the data obtained. It is obvious, however, that many more data need to be gathered, before rational clinical decisions can be based on Xe-CT CBF studies. A more general problem in this area is to develop systems for consistent storage and analysis of the abundance of obtained data, in such way that meaningful and clinically useful information can be extracted from it. Ideally, such databases must be designed prospectively with clearly formulated clinical questions in mind.

References

1. Abdel-Dayem HM, Sadek SA, Kouris K, Bahar RH, Higazi I, Eriksson S, Englesson SH, Berntman L, Sigurdsson GH, Foad M, Olivecrona H (1987) Changes in cerebral perfusion after acute head injury: comparison of CT with Tc-99m HM-PAO SPECT. Radiology 165: 221–226
2. Alavi A, Weller S, Alves WM, Spielman GM, Reivich M, Gennarelli TA (1989) Distinct patterns of metabolic changes in head injury as detected by positron emission tomography. J Cereb Blood Flow Metab 9 [Suppl 1]: S379

3. Becker DP, Gade GF, Young HF, Feuerman TF (1990) Diagnosis and treatment of head injury in adults. In: Youmans JR (ed) Neurological surgery. W. B. Saunders Co., Philadelphia, pp 2017–2148

4. Bouma GJ, Muizelaar JP, Choi SC, Newlon PG, Young HF (1991) Cerebral circulation and metabolism after severe traumatic brain injury: the elusive role of ischemia. J Neurol Neurosurg 75: 685–693

5. Bouma GJ, Muizelaar JP, Stringer WA, Choi SC, Fatouros PP, Young HF (1992) Ultra-early evaluation of regional cerebral blood flow in severely head-injured patients using xenon-enhanced computerized tomography. J Neurosurg 77: 360–368

6. Cold GE (1990) Cerebral blood flow in acute head injury. The regulation of cerebral blood flow and metabolism during the acute phase of head injury, and its significance for therapy. Acta Neurochir (Wien) [Suppl] 49: 1–64

7. Darby JM, Yonas H, Pentheny SL, Marion DW (1989) Intracranial pressure response to stable xenon inhalation in patients with head injury. Surg Neurol 32: 343–345

8. Drayer BP, Wolfson SK, Reinmuth OM, Dujovny M, Boehnke M, Cook EE (1978) Xenon-enhanced CT for analysis of cerebral integrity, perfusion and blood flow. Stroke 9: 123–130

9. Enevoldsen EM (1986) CBF in head injury. Acta Neurochir (Wien) [Suppl] 36: 133–136

10. Fatouros PP, Wist AO, Kishore PR, DeWitt DS, Hall JA, Keenan RL, Stewart LM, Marmarou A, Choi SC, Kontos HA (1987) Xenon/computed tomography cerebral blood flow measurements: methods and accuracy. Invest Radiol 22: 705–712

11. Fieschi C, Battistini N, Beduschi A, Boselli L, Rossanda M (1974) Regional cerebral blood flow and intraventricular pressure in acute head injuries. J Neurol Neurosurg Psychiatry 37: 1378–1388

12. Good WF, Gur D (1987) The effect of computed tomography noise and tissue heterogeneity on cerebral blood flow determination by xenon-enhanced computed tomography. Med Phys 14: 557–561

13. Gur D, Wolfson SK, Yonas H, Good WF, Shabason L, Latchaw RE, Miller DM, Cook EE (1982) Progress in cerebrovascular disease: local cerebral blood flow by xenon enhanced CT. Stroke 13: 750–758

14. Gur D, Yonas H, Good WF (1989) Local cerebral blood flow by xenon-enhanced CT: current status, potential improvements, and future directions. Cerebrovasc Brain Metab Rev 1: 68–86

15. Harrington TR, Manwaring K, Hodak J (1986) Local basal ganglia and brain stem blood flow in the head-injured patient using stable xenon enhanced CT scanning. In: Miller JD, Teasdale GM, Rowan JO, et al (eds) Intracranial pressure, Vol 6. Springer, Berlin Heidelberg New York Tokyo, pp 680–686

16. Hughes RL, Yonas H, Gur D, Latchaw RE (1989) Cerebral blood flow determination within the first 8 hours of cerebral infarction using stable xenon-enhanced computed tomography. Stroke 20: 754–760

17. Langfitt TW, Obrist WD (1981) Cerebral blood flow and metabolism after intracranial trauma. In: Krayenbühl H (ed) Progress in neurological surgery, Vol 10. S. Karger, Basel, pp 14–48

18. Latchaw RE, Yonas H, Darby JM, Pentheny SL (1986) Xenon/CT cerebral blood flow determination following cranial trauma. Acta Radiol [Suppl] 369: 370–373

19. Latchaw RE, Yonas H, Pentheny SL, Gur D (1987) Adverse reactions to xenon-enhanced CT cerebral blood flow determination. Radiology 163: 251–254

20. Marion DW, Bouma GJ (1991) The use of stable Xenon CT CBF studies to define changes in cerebral CO_2 vasoresponsivity caused by severe head injury. Neurosurgery 29: 869–873

21. Marion DW, Crosby K (1991) The effect of stable xenon on ICP. J Cereb Blood Flow Metab 11: 347–350

22. Marion DW, Darby JM, Yonas H (1991) Acute regional cerebral blood flow changes caused by severe head injuries. J Neurosurg 74: 407–414

23. Marmarou A, Anderson RL, Ward JD, Choi SC, Young HF, Eisenberg HM, Foulkes MA, Marshall LF, Jane JA (1991) Impact of ICP instability and hypotension on outcome in patients with severe head trauma. J Neurosurg 75: S59–S66

24. Marshall LF (1990) Current head injury research. Curr Opin Neurol Neurosurg 3: 4–9

25. Meyer JS, Hayman LA, Yamamoto M, Sakai F, Nakajima S (1980) Local cerebral blood flow measured by CT after stable xenon inhalation. AJNR 1: 213–215

26. Miller JD (1982) Disorders of cerebral blood flow and intracranial pressure after head injury. Clin Neurosurg 29: 162–173

27. Miller JD, Becker DP, Ward JD, Sullivan HG, Adams WE, Rosner MJ (1977) Significance of intracranial hypertension in severe head injury. J Neurosurg 47: 503–515

28. Muizelaar JP, Obrist WD (1985) Cerebral blood flow and brain metabolism with brain injury. In: Becker DP, Povlishock JT (eds) Central nervous system trauma status report. National Institute of Health, Bethesda, pp 123–137

29. Obrist WD, Langfitt TW, Jaggi JL, Cruz J, Gennarelli TA (1984) Cerebral blood flow and metabolism in comatose patients with acute head injury. J Neurosurg 61: 241–253

30. Obrist WD, Wilkinson WE (1979) The non-invasive Xe-133 method: Evaluation of cerebral blood flow indices. In: Bes A, Geraud G (eds) Cerebral circulation. Excerpta Medica Elsevier, Amsterdam, pp 119–124

31. Obrist WD, Wilkinson WE (1990) Regional cerebral blood flow measurement in humans by xenon-133 clearance. Cerebrovasc Brain Metab Rev 2: 283–327

32. Overgaard J, Tweed WA (1974) Cerebral circulation after head injury. Part 1: CBF and its regulation after closed head injury with emphasis on clinical correlations. J Neurosurg 41: 531–541

33. Pistolese GR, Faraglia V, Agnoli A, Prencipe M, Pastore E, Spartera C, Fiorani P (1972) Cerebral hemispheric "countersteal" phenomenon during hyperventilation in cerebrovascular diseases. Stroke 3: 456–461

34. Segawa H (1985) Tomographic cerebral blood flow measurement using xenon inhalation and serial CT scanning: normal values and its validity. Neurosurg Rev 8: 27–33

35. Siesjö BK, Wieloch T (1985) Brain injury: neurochemical aspects. In: Becker DP, Povlishock JT (eds) Central nervous system trauma status report – 1985. National Institute of Health, Bethesda, pp 513–532

36. Stringer WA, Marion DW, Bouma GJ, Muizelaar JP, Braun IF, Fatouros PP, Marmarou A (1992) Correlation of xenon-133 and stable xenon-enhanced computed tomographic cerebral blood flow measurements in patients with severe head injury. In: Yonas H (ed) Cerebral blood flow measurement with stable xenon-enhanced computed tomography. Raven Press, New York, pp 233–237

37. Tenjin H, Ueda S, Mizukawa N, Imahori Y, Hino A, Yamaki T, Kuboyama T, Ebisu T, Hirakawa K (1990) Positron

emission tomographic studies on cerebral hemodynamics in patients with cerebral contusion. Neurosurgery 26: 971–979

38. Winkler S, Sacket J, Holden J, Fleming DC, Alexander SC, Madsen M, Kimmel RI (1977) Xenon inhalation as an adjunct to computerized tomography of the brain: Preliminary study. Invest Radiol 12: 15–18

39. Winkler S, Turski P (1985) Potential hazards of xenon inhalation. AJNR 6: 974–975

40. Wolfson SK, Clark J, Greenberg JH, Gur D, Yonas H, Brenner RP, Cook EE, Lordeon PA (1990) Xenon-enhanced computed tomography compared with [14-C]-Iodoantipyrine for normal and low cerebral blood flow states in baboons. Stroke 21: 751–757

41. Yonas H, Wolfson SK, Gur D, Latchaw RE, Good WF, Leanza R, Jackson DL, Jannetta PJ, Reinmuth OM (1984) Clinical experience with the use of xenon-enhanced CT blood flow mapping in cerebral vascular disease. Stroke 15: 443–450

Correspondence and Reprints: Gerrit J. Bouma, M. D., Department of Neurosurgery, University of Amsterdam, Academic Medical Center, 1105 AZ Amsterdam, The Netherlands.

Continuous Monitoring of Cerebral Blood Flow
and Metabolism in Intensive Care

Acta Neurochir (1993) [Suppl] 59: 43–46

Thermodiffusion

Cerebral Blood Flow (CBF) Monitoring in Intensive Care by Thermal Diffusion

L. P. Carter[1], **M. E. Weinand**[1], and **K. J. Oommen**[2]

[1] Section of Neurosurgery and [2] Department of Neurology, University of Arizona School of Medicine, Tuscon, Arizona, U.S.A.

Summary

Continuous monitoring of cortical blood flow (CoBF) in the intensive care unit is possible with thermal diffusion techniques. The normal brain flow limits have been established when electrical activity ceases and when infarction is likely to occur. With continuous monitoring of CoBF one can see immediate changes in flow and approaching these levels may be anticipated.

The thermal diffusion system we have employed is based on the thermal conductivity of cortical tissue. As blood flow increases through the tissue, the conduction of energy away from the flow probe allows the sensor to detect changes in flow.

This form of monitoring has been carried out in patients with subarachnoid hemorrhage, resection of cerebral mass lesions, severe craniotrauma, and intractable epilepsy. In subarachnoid hemorrhage, vasospasm can be identified and the efficacy of treatment determined with continuous monitoring of CoBF. During resection of mass lesions, increases in blood flow can be readily detected to document the recovery of brain tissue. Continuous monitoring of CoBF in epilepsy patients is now possible with the implantation of subdural electrodes. The increase in blood flow can be documented and it is apparent that a period of elevation of blood flow is quite short. Therefore, this may be helpful in determining when other forms of CBF determination, such as Single Photon Emission Computed Tomographic (SPECT) scanning should be performed. In patients with cranial trauma, different patterns of CoBF changes are apparent. Some patients may develop increased CoBF prior to elevation of intracranial pressure (ICP); other patients demonstrate a drop in CoBF as a response to increased ICP. It is apparent that by knowing CoBF, these patients may be treated more rationally, either by changing their ventilation or by suppressing brain metabolism.

Further studies are necessary to define its true value in overall patient management.

Keywords: Cortical blood flow; thermodiffusion; intensive care; continuous monitoring.

Introduction

The brain is the most abundantly perfused organ in the body, receiving fully 20% of the cardiac output for only approximately 1 500 g of tissue. Normal whole brain blood flow runs 50 ml/100 g of tissue/minute while gray matter flow is in the order of 70 ml/g of tissue/minute. Supranormal flows occur in such conditions as hypertensive encephalopathy, epileptogenic seizure discharge, and normal perfusion pressure breakthrough (NPPB). As blood flow drops, protein synthesis ceases below 40 ml/100 g tissue/minute[10]. Spontaneous electrical activity stops at 20 ml/100 g of tissue/minute[1]. The evoked responses fall off at approximately 15 ml/100 g of tissue/minute, and the potassium/sodium ion pump fails at 8–10 ml/100 g of tissue/minute[1]. Neuronal survival depends upon the duration of the reduction of CBF. At very low flows neurons can survive for only 5 to 10 minutes. However, at high flows these neurons are capable of surviving for longer periods of time. We have demonstrated a time duration curve using the direct cortical response as a physiological marker of neural integrity and have found that at 5 ml/100 g of tissue/minute, the cortical tissue can recover for up to 30 minutes; whereas at 10 ml/100 g of tissue/minute, the cortex can regain function for up to 50–60 minutes[9]. At higher levels, this time period is further prolonged.

Under conditions of acute brain injury such as trauma or subarachnoid hemorrhage, the brain's normal autoregulatory mechanism may be impaired. Therefore, changes in perfusion pressure either by a elevation of ICP or reduction of systemic blood pressure may compromise CBF. ICP is frequently monitored in critical care patients as an indirect measure of perfusion. However, it may be more important to look at the actual flow through the tissue since this can now be monitored effectively.

Methods

We have employed a Flowtronics Saber (Flowtronics, Inc., Phoenix, AZ) thermodiffusion blood flow system which gives a continuous, real-time evaluation of CoBF. The sensor has two gold plates, one heated and one neutral, and the temperature difference between these two plates is constantly measured. At 0 blood flow, the temperature difference (Δ T) is known since the cortical tissue has a fixed conductivity constant. As the blood flow increases through this capillary bed, the heated plate looses heat by conduction from the blood flow. As this blood flow increases, the heated plate comes closer to the neutral plate and ΔT becomes smaller. ΔT has been correlated with the cortical blood flow in ml/100 g tissue/minute. This was initially described by comparing Xenon 133, and subsequently with hydrogen clearance[5]. A recent correlation by Robertson, in 10 patients, comparing CoBF with the thermodiffusion technique to the Kety-Schmidt nitrous oxide method, demonstrated an excellent correlation between the two techniques despite the fact that the nitrous oxide method measures whole brain CBF, whereas the thermodiffusion system measures regional flow in the cortex[6] only (see Fig. 1).

The sensor is connected to a microprocessor which continuously converts the temperature difference to ml/100 g of tissue/ minute. Acute changes in flow are readily apparent. Some examples in the Operating Room of a temporary clip being applied to a major cerebral vessel as well as abrupt changes in perfusion pressure occurring in laboratory animals have defined the rapid

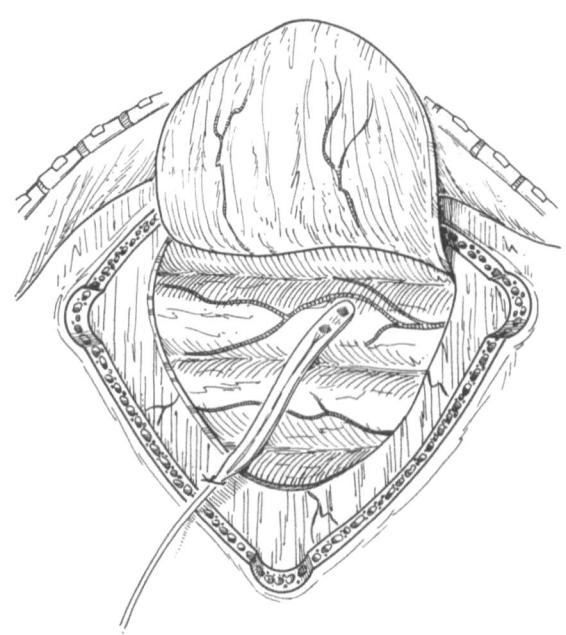

Fig. 2. Thermodiffusion sensor placed on normal cortex away from large surface vessels

response time of this system[3]. The thermal sensor must rest on the cortex. If the thermal gradient is diluted by blood, cerebral spinal fluid, or air, a false reading will be obtained. To determine whether the sensors are in contact with the cortex, a confidence check has been developed. Under these circumstances, the heater is discontinued and the two plates come to near equal temperatures. If they are both in touch with the same tissue such as the cortex, the confidence factor will be greater than 50 and will be considered reliable. The sensor must be placed through a burr hole or craniotomy site and is placed on an area of cortex of interest. Normal cortex is chosen, an attempt is made to avoid large surface vessels, areas of contusion, or other pathological tissue (see Fig. 2).

Results and Discussion

A probe is placed either at the time of surgery or through a burr hole in a wide variety of neurological conditions such as subarachnoid hemorrhage, mass lesions, epilepsy, and trauma cases.

In subarachnoid hemorrhage patients with intracranial aneurysms, a significant cause of morbidity and mortality is vasospasm. By monitoring CoBF, one can detect alterations in flow or flow reductions which occur with vasospasm[2]. The transcranial Doppler only gives the flow velocity within the basal arteries and is generally done intermittently. A continuous recording of CoBF can allow detection of vasospasm as well as judging treatment[2]. If the treatment is to improve CBF by raising the blood pressure

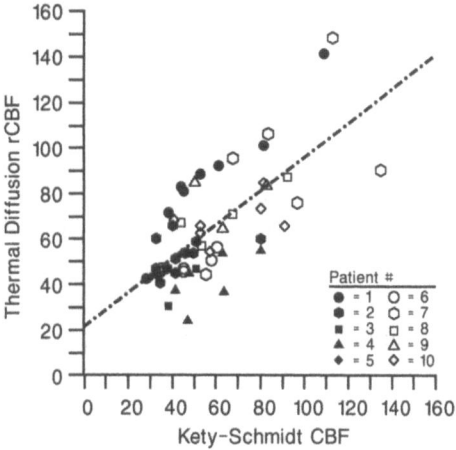

Fig. 1. Comparison of CoBF as determined by thermodiffusion with Kety-Schmidt CBF measurements in 10 patients. Courtesy of C. Robertson, Baylor College of Medicine, Houston, Texas

Day 2

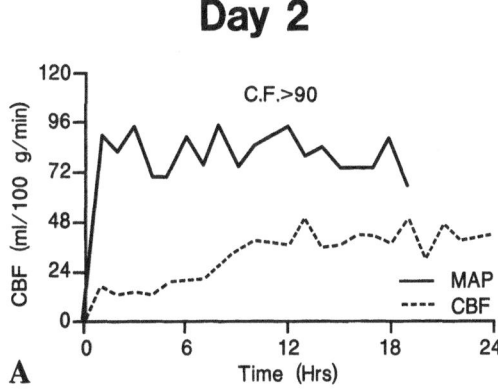

Day 3

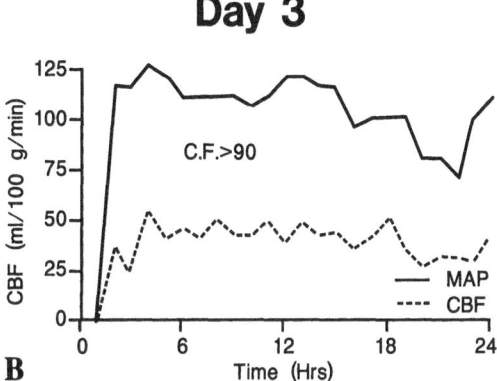

Fig. 3. Monitor of BP and CoBF in patient with severe vasospasm. (A) Day 2, post-op. (B) Day 3, post-op. BP was elevated in order to maintain adequate CoBF. Patient made an excellent recovery. Note improved CoBF with sustained hypertension

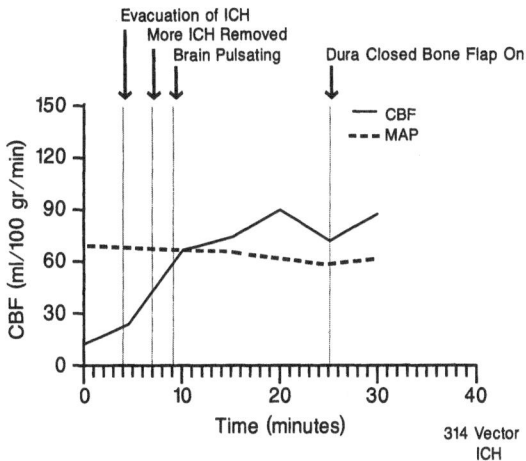

Fig. 4. CoBF during evacuation of intracerebral hematoma. CoBF was low prior to removal of mass. After the hematoma was removed, CoBF goes to hyperemic levels

have measured flow in and around benign and malignant brain tumors and have demonstrated that the area of low density around the tumor frequently has markedly reduced flows. This flow often increased to hyperemic levels after tumor resection. Such a marked elevation in CoBF may be a prelude to malignant cerebral edema.

It is of interest that with intractable epilepsy, blood flows in the interictal focus are found to be markedly reduced; whereas, during the seizure discharge, we see a significant elevation of CoBF (Fig. 5). We are currently investigating this technique to see if measurement of flow can aid in the localization of the epileptogenic focus[9].

Perhaps the most helpful area of monitoring CoBF is in severe cranial trauma. There is a tendency at most institutions to immediately hyper-

or increasing the volume, then one can assess the effectiveness of the treatment (see Fig. 3).

Patients with arteriovenous malformations may develop problems with NPPB[7]. This occurs after the shunt has been relieved and more blood is traversing through the normal tissue. As blood flow increases, the hyperemic tissue may become edematous and hemorrhagic leading to a life-threatening situation. By monitoring CoBF in areas around AVM resection, we have consistently found the blood flow to increase following resection and flow can be brought under control by reducing blood pressure with barbiturates.

After resection of cerebral mass lesions such as hematomas or tumors, there is a significant rise in CoBF presumably due to the release of local cortical capillary compression. An example of this is in the evacuation of an intracerebral hematoma in which the CoBF is quite low prior to evacuation, with marked increase in CoBF following this (Fig. 4). We

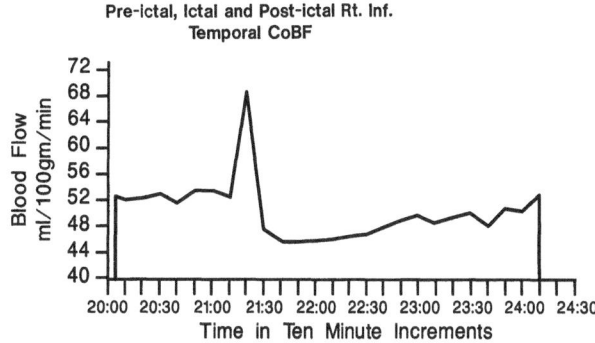

Fig. 5. Monitoring CoBF during an epileptic seizure. Note the marked elevation with seizure discharge and then depression in CoBF after seizure

ventilate the patient and treat them with mannitol in a "shotgun" technique. Both of these maneuvers will theoretically lower ICP. However, patients who already have a reduced CBF may not be good candidates for hyperventilation and patients with elevated CBF should probably not receive mannitol as the initial form of treatment. By monitoring patients with cranial trauma and, particularly, patients postoperatively from subdural hematoma evacuation, we frequently see elevations in CoBF precede elevations of ICP[2, 4]. It is apparent that we can more rationally treat these patients by trying to maintain adequate CBF as opposed to hyperventilating all patients. One of the thermodiffusion flow probes has an ICP port so that both ICP and CoBF can be continuously monitored during the course of the patient's care. Whether this type of manipulation of the CoBF can improve overall patient outcome can only be surmised after evaluating a large group of patients, probably in a multi-center trial.

In general, continuous monitoring of CoBF with the thermodiffusion technique is safe. We have had only one patient who developed a significant infection with this type of monitoring, and that patient also had a ventriculostomy. An additional epilepsy patient developed aseptic meningitis which was treated with steroids. This is a novel method of evaluating cortical perfusion and may aid in the treatment of patients with vasospasm, mass lesions, epilepsy, and cranial trauma. Further studies are

necessary to define its true usefulness in overall patient management.

References

1. Astrup J, Symon L, Branston NM, *et al* (1977) Cortical evoked potential and extracellular K^+ and H^+ at critical levels of brain ischemic. Stroke 8: 51–57
2. Carter LP (1991) Surface monitoring of cerebral cortical blood flow. Cerebrovasc Brain Metab Rev 3: 246–261
3. Carter LP, Erspaner JR, White WL, *et al* (1982) Cortical blood flow during craniotomy for aneurysms. Surg Neurol 17: 204–208
4. Dickman CA, Carter LP, Baldwin HZ, *et al* (1991) Technical report. Continuous regional cerebral blood flow monitoring in acute craniocerebral trauma. Neurosurgery 28: 467–472
5. Gaines C, Carter LP, Crowell RM (1983) Comparison of local cerebral blood flow determined by thermal and hydrogen clearance. Stroke 14: 66–69
6. Robertson C (1993) Baylor College School of Medicine, Houston, Texas (personal communication)
7. Spetzler RF, Wilson CB, Weinstein P (1978) Normal perfusion pressure breakthrough theory. Clin Neurosurg 25: 651–652
8. Weinand ME, Carter LP, Oommen KJ, *et al* (1992) Surface monitoring of cerebral cortical blood flow in epilepsy. Neurology 42 [Suppl 3]: 81
9. Yamagata S, Carter LP, Erspamer RJ (1982) Cortical ischemia effect upon direct cortical response. Stroke 13: 680–686
10. Xie Y, Mies G, Hossman KA (1984) Ischemic threshold of brain protein synthesis of the unilateral carotid artery occlusion in gerbils. Stroke 20: 620–626

Correspondence and Reprints: L. P. Carter, M. D., Section of Neurosurgery, University of Arizona School of Medicine, Tucson, AZ 85724, U.S.A.

Acta Neurochir (1993) [Suppl] 59: 47–49

Monitoring of Regional Cerebral Blood Flow (CBF) in Acute Head Injury by Thermal Diffusion

M. L. Schröder and **J. P. Muizelaar**

Division of Neurosurgery, Medical College of Virginia, Virginia Commonwealth University, Richmond, Virginia, U.S.A.

Summary

During the last few years continuous measurements of CBF by means of a thermal diffusion blood flow probe have been proposed as a possible means for monitoring the patient's CBF in a clinical setting. Also, it has been suggested that continuous CBF data from head injured patients can be correlated with other continuously recorded clinical parameters, such as ICP and blood pressure, in order to clarify pathophysiological mechanisms such as "plateau-waves".

We measured regional cortical blood flow continuously with a thermal diffusion flow probe in 13 comatose head injured patients after undergoing craniotomy for evacuation of a traumatic intracranial mass lesion in order to assess the reliability and usefulness of the method. In seven patients stable Xenon-CT CBF studies were performed with the flow probe in place, in order to compare the two methods. The continuous blood flow values did not correlate with regional or global stable Xenon-CT values. These results indicate that continuous monitoring of CBF with the thermal diffusion method as currently used cannot be used in the clinical management of the patient. Further research will have to be directed to the question as to whether changes in CBF are reliably measured with this method. If this is true, the thermal diffusion flow probe with its high temporal resolution may still be useful in investigating pathophysiological mechanisms such as interaction between CBF, ICP, mean arterial blood pressure (MABP), and endexpiratory CO_2 (etCO$_2$).

Keywords: Cerebral blood flow; thermal diffusion; stable Xenon-CT.

Introduction

Thermal diffusion flow probes have been shown to reliably indicate the level of CBF in a number of conditions, both experimental and clinical[2, 4, 7]. In 1991 Dickman and co-workers applied this method for the first time in head injured patients. These authors noted that this method could yield data relating to pathophysiological mechanisms but "this method of CBF monitoring has not yet been established for clinical decision making"[5]. However, if it could be shown that the CBF values obtained with this method correlate well with those obtained with the stable Xenon-CT method[1, 9], then continuous monitoring of CBF with the thermal diffusion blood flow probe could be applied in a clinical setting.

Material and Methods

In thirteen head injured patients with a mean Glasgow Coma Score of 6 (range:3–10) a thermal diffusion blood flow sensor (Micro Saber Plus Blood Flow Monitoring System, Flowtronics, Phoenix, Arizona; CBF sensor model: no. MS 7000)[3, 8] was inserted during craniotomy performed for treatment of an intracranial mass lesion following published procedural guidelines[6]. The plates of the probe were consistently placed in the operation field and moved up slightly under the dura to obtain good contact with the brain tissue. Next, a "confidence check" was done to ensure reliable recording (Flowtronics, unpublished). Upon admission to the intensive care unit CBF-, ICP-, MABP-, and etCO$_2$-values were simultaneously recorded every 30 seconds and entered into a database in a VAX computer. The age of these patients varied between 15 and 82 years (median: 27), and 12 (92%) were males. Three patients had epidural hematomas, eight had subdural hematomas, and two had intracerebral hematomas. ICP was measured with a ventriculostomy. Seven patients underwent one or more stable Xenon-CT CBF studies as part of a postoperative control CT scan or follow-up CT scan within 5 days after injury to correlate the two methods. To compare probe values with the stable Xenon-CT values a mean cortical blood flow value was determined for the cerebral area adjacent to the flow probe by using circular regions of interest of 2 cm diameter (Fig. 1). In four patients we also used a 3 to 5 mm region of interest contiguous to the probe. During these CBF studies continuous blood flow probe data were collected with a portable unit, consisting of a data recording system (Maclab 8™ data recording system, Castle Hill, Australia) and amplifier connected to a Macintosh computer (Apple Computer, Inc. Cupertino, California). The stable Xenon-CT CBF studies were performed on a GE-9800 scanner (General Electric, Milwaukee, Wisconsin).

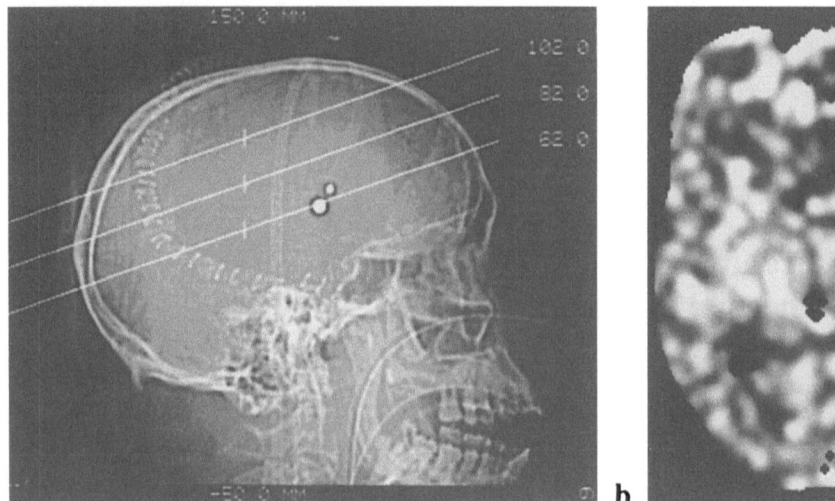

a   b

Fig. 1. Scout view (a) and corresponding stable Xenon-CT image (b) of a patient with a large hemorrhagic contusion in the left frontal lobe shortly after injury. Note the probe placement close to the contusion and the mean blood flow in a 2 cm region adjacent to the tip of the sensor

Results

The mean duration of the probe placement was 98 hours (range: 1 to 7.5 days). Two patients had a positive probe-tip culture and one of them developed an intracerebral abscess requiring craniectomy. Two patients had a positive craniotomy wound culture and one of them developed an infected boneflap requiring surgery. Two patients had a CSF leak out of the stab wound for the probe and one of them developed meningitis. Limited data were collected in 6 cases: In one case the patient expired soon after admission to the I.C.U., one patient started obeying commands, in one case the cable broke and in three patients CBF constantly varied between O and 200 ml/100 g/min without obvious reason and the data were deemed unreliable. The stable Xenon-CT data from the 2 cm and the 3–5 mm regions of interest adjacent to the flow probe were found to have no correlation with the probe measurements (r = 0.57, p = 0.17 and r = 0.53, p = 0.44 respectively; regression analysis). Similarly, no correlation was established when comparing the probe data to the global hemisheric Stable Xenon-CT CBF values. The data obtained by the two CBF methods is shown in Fig. 2.

Discussion

This study demonstrates that the use of the current model of the thermal diffusion blood flow probe

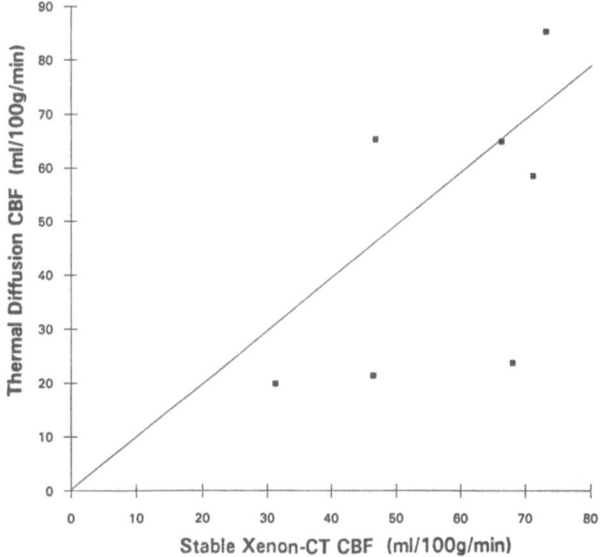

Fig. 2. Graph showing the relationship between CBF values obtained with the stable Xenon-CT method and the thermal diffusion blood flow sensor. No significant relationship could be found between the two methods (r = 0.57, p = 0.17)

is associated with complications and does not provide dependable CBF values. Although some linear relationship between the two methods could be distinguished, the highly significant correlation necessary for application of the thermal diffusion blood flow probe in patients' management could not be obtained. Therefore, at the present time measure-

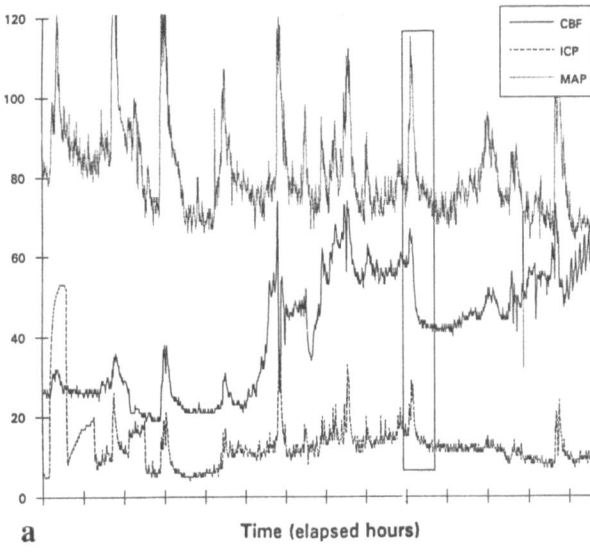

a   Time (elapsed hours)

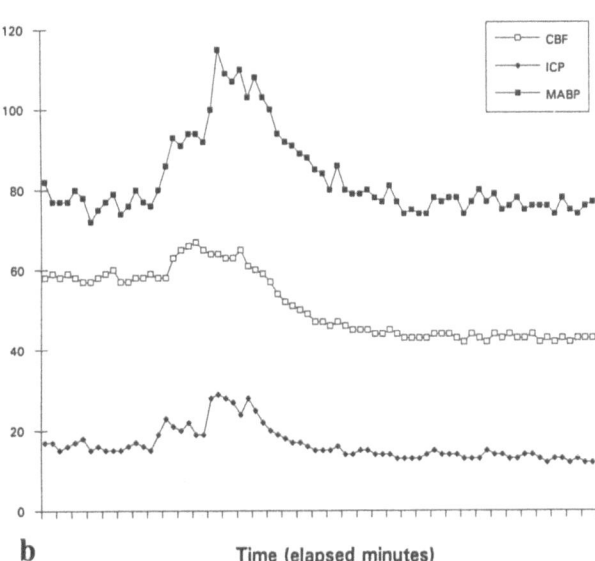

b   Time (elapsed minutes)

Fig. 3. (a) Simultaneous CBF, ICP, and MABP recordings with a 30 second resolution in a severely head injured patient; (b) represents an expanded time interval indicated by the rectangle in (a). CBF = ml/100 g/min, ICP = mm Hg, MABP = mm Hg

mining synchronous changes of different physiological parameters: CBF, ICP, MABP, and etCO$_2$[7] (Fig. 3). Recently, we are able to monitor these parameters with a higher time resolution (5 data/sec) which allows more precise detection of abrupt changes. Moreover, the results could possibly be improved by using a more advanced probe (CBF sensor model: no.M 9800, Flowtronics, Phoenix, Arizona) recently available. In summary, the probe as presently configured is not suitable for clinical use and further tests are necessary to establish the utility of newer probe designs.

References

1. Bouma GJ, Muizelaar JP (1992) Cerebral blood flow, cerebral blood volume, and cerebrovascular reactivity after severe head injury. J Neurotrauma 9 [Suppl 1]: 333–348
2. Carter LP, Atkinson JR (1973) Cortical blood flow in controlled hypotension as measured by thermal diffusion. J Neurol Neurosurg Psychiatry 36: 906–913
3. Carter LP, Erspamer R, Bro WJ (1981) Cortical blood flow: Thermal diffusion vs isotope clearance. Stroke 12: 513–518
4. Carter LP, Grahm T, Bailes JE, Bichard W, Spetzler RF (1991) Continuous postoperative monitoring of cortical blood flow and intracranial pressure. Surg Neurol 35: 36–39
5. Dickman CA, Carter LP, Baldwin HZ, Harrington T, Tallman D (1991) Continuous regional cerebral blood flow monitoring in acute craniocerebral trauma. Neurosurgery 28: 467–472
6. Ohmoto T, Nagao S, Mino S, Fujiwara T, Honma Y, Ito T, Ohkawa M (1991) Monitoring of cortical blood flow during temporary arterial occlusion in aneurysm surgery by the thermal diffusion method. Neurosurgery 28: 49–54
7. Rosner MJ, Becker DP (1984) Origin and evolution of plateau waves. Experimental observations and a theoretical model. J Neurosurg 50: 312–324
8. Salcman M, Moriyama E, Elsner HJ, Rossman H, Gettleman RA, Neuberth G, Corradino G (1989) Cerebral blood flow and the thermal properties of the brain: a preliminary analysis. J Neurosurg 70: 592–598
9. Yonas H (1985) Measurement of cerebral blood flow. In: Wilkins RH, Rengachary SS (eds) Neurosurgery. McGraw-Hill, New York, pp 1173–1178

ments of CBF by means of this probe are not recommended for clinical use. However, this probe could be used in detecting and observing physiological mechanisms in an experimental setting such as deter-

Correspondence and Reprints: J. P. Muizelaar M. D., Ph.D., Division of Neurosurgery, Medical College of Virginia, MCV Station Box 631, Richmond, VA 23298-0631, U.S.A.

Acta Neurochir (1993) [Suppl] 59: 50–57

Tissue pO$_2$

Monitoring Cerebral Oxygenation: Experimental Studies and Preliminary Clinical Results of Continuous Monitoring of Cerebrospinal Fluid and Brain Tissue Oxygen Tension

A. I. R. Maas, W. Fleckenstein, D. A. de Jong, and **H. van Santbrink**

Departments of Neurosurgery and Experimental Surgery, Academic Hospital, Erasmus University Rotterdam, The Netherlands

Summary

Cerebral ischaemia is considered to be the central mechanism leading to secondary brain damage in patients with severe head injury. It would therefore seem appropriate to monitor cerebral oxygenation in these patients. The possibilities of continuous monitoring of brain tissue and CSF oxygen tension as parameters for cerebral oxygenation were evaluated. In experimental studies the influence of changed oxygen offer and decreased cerebral perfusion pressure on CSF and brain tissue pO$_2$ were investigated. Fast changes in CSF pO$_2$ were observed in response to decreasing oxygen offer. Slower changes were noted in response to hypo- and hyperventilation. An autoregulatory mechanism regulating CSF pO$_2$ is postulated. Reducing cerebral perfusion pressure decreased both brain tissue and CSF pO$_2$, but in the reperfusion phase after complete ischaemia a dissociation occurred between brain tissue and CSF pO$_2$, CSF pO$_2$ being restored, but brain tissue pO$_2$ remaining low or even decreasing further. From these studies it is concluded that both CSF pO$_2$ and brain tissue pO$_2$ reflect changes in cerebral oxygenation caused by changes in oxygen offer as well as by changes in cerebral blood flow. Brain tissue pO$_2$ is also sensitive to oxygen demand from the tissue.

Preliminary studies of continuous monitoring of brain tissue pO$_2$ in patients with severe head injury are reported.

Keywords: Cerebral oxygenation; oxygen tension monitoring; CSF pO$_2$; brain tissue pO$_2$; severe head injury; ischaemia.

Introduction

Monitoring intracranial pressure (ICP) and treatment of raised ICP is generally accepted as state of the art care in patients with severe head injury. Yet the question has been raised whether increased ICP merely reflects the severity of brain damage, or is indeed a further cause of secondary brain injury.

Recently attention has been focussed on calculating cerebral perfusion pressure (CPP) as a guideline to management in these patients[8, 11]. This is the result of the realisation that the main mechanism of secondary brain injury due to raised ICP is caused by decreased cerebral blood flow (CBF) and oxygen offer to the brain tissue. Ischaemia is considered to be the central mechanism leading to secondary brain damage in patients with severe head injury. At autopsy ischaemic changes have been noted in more than 90% of patients dying due to the head injury[7].

Cerebral ischaemia is caused by both extracerebral factors (arterial hypoxia) and general and regional changes in cerebral blood flow. It would therefore seem appropriate to monitor cerebral oxygenation. Cerebral oxygenation is determined by the oxygen offer to the brain, the cerebral blood flow and the rate of cerebral oxygen metabolism. For clinical practice the monitoring technique employed should reflect the net result of these determinants and also yield continuous measurements. With these considerations in mind we have investigated the possibilities of continuous monitoring of brain tissue- and CSF oxygen tension (pO$_2$) as parameters for cerebral oxygenation. Basic aspects concerning the technique of measurement and the influence of changing ventilatory conditions, as well as the effect of raised ICP and decreased CPP were evaluated in animal experiments.

The feasibility of the technique for routine clinical practice was further evaluated in preliminary studies in patients with severe head injury.

Material and Methods

Experimental Studies: Reference Conditions and Ventilatory Experiments

In six cats (weight 2.5–4.0 kg) and five beagle dogs (weight 12.3–22.5 kg) anaesthesia was induced with ketamine (40 mg in cats) or methohexital (3 mg/kg in dogs) and maintained by continuous infusion of fentanyl (0.03 mg/kg/h in cats; 0.015 mg/kg/h in dogs) and pancuronium (0.05 mg/kg/h in cats; 0.08 mg/kg/h in dogs) and an admixture of N_2O to the respired gas. The animals were artificially ventilated. Rectal temperatures were stabilized between 36.4° and 38.2°C. Catheters were placed in the femoral artery, femoral vein and in the urinary bladder. Subsequently, in order to install brain sensors the frontoparietal bones were exposed and avoiding CSF leakage two small holes were drilled through the skull (diameter 1 mm for oxygen needle probe and thermocouple; 1.5 mm for ventricular catheter). Through these burrholes a needle was positioned in the right cella media for continuous monitoring of the ventricular fluid pressure and the pO_2 sensor and thermocouple were positioned in the cella media of the left lateral ventricle. In the cat experiments a second pO_2 probe was positioned in the cisterna magna for continuous recording of cisternal CSF pO_2. Arterial blood pressure, ECG, heart rate, ventricular CSF pO_2, in the cats also cisternal CSF pO_2, brain temperature and rectal temperature were continuously monitored.

In order to record "reference" CSF pO_2 values, measurements were started one hour after termination of the operative procedures at normocapnic ventilation and slightly elevated fractional volume of inspired O_2 (FIO$_2$ of 0.25). The reference CSF pO_2 values were recorded in the subarachnoid space during positioning of the sensor and further continuously in ventricular and cisternal CSF. Subsequently the animals were hypo- and hyperventilated and subjected to hypercarbic, hypoxic and hyperoxic ventilation. Each of the experiments (duration 6–20 min) was ended by re-installing reference conditions.

Experimental Studies: Effect of Increased ICP and Decreased CPP on Cerebral Oxygenation

The studies were performed in 10 dogs, all anaesthetized and artificially ventilated. The experimental set-up was the same as described for the reference conditions and ventilatory experiments. The arterial pO_2 and PCO_2 were kept within narrow limits during these studies at mean values of 110 and 38 mmHg

respectively. In these experiments the brain tissue pO_2 was also continuously recorded in addition to the ventricular CSF pO_2.

Intracranial pressure was raised by cisternal infusion of Ringers lactate. Intracranial pressure was increased stepwise by increments of 10 mmHg, each sustained for 15 min. A problem resulted from increasing arterial blood pressure secondary to raised ICP. This so-called "Cushing response" could not adequately be prevented by pharmacological agents, but was effectively inhibited by inducing on artificial pericardiac tamponade in six animals.

Technique of Continuous CSF pO_2 Measurements

CSF pO_2 was measured with polarographic needle probes (low sensitive "Licox" probe type, GMS mbH Kiel, Federal Republic of Germany). The outer tube of the probes is made of spring steel. pO_2 is measured at a membranized gold microcathode, situated within a recess of the probe tip (goldwire diameter: 0.0125 mm; goldwire installation: glass sealing; reaction time T90 of the probe current after pO_2 change: <2 seconds). Sensitivity drifting: <2%/h; no measurable PH sensitivity between PH 6.5 and 9.5; before and after insertion in CSF the probes were calibrated in physiological saline solution, saturated either with nitrogen or with air. The probe current in CSF (measured in intervals of 4 seconds) was stored electronically thus allowing recalibration calculations (device "Licox" GMS mbH Kiel, Federal Republic of Germany). Temperature compensation was performed automatically every 20 seconds.

Brain Tissue pO_2 Measurements

Brain tissue pO_2 was measured in the experimental studies on CPP and in six patients with severe head injury continuously from a small flexible polarographic (Clark type) catheter placed in the right frontal region. The measuring device used was the Licox pO_2 monitor (manufactured by GMS mbH Kiel). The pO_2 catheter was introduced through a vein catheter needle (Abbocath) and positioned in the white matter. Local tissue pO_2 values were integrated from a tissue cylinder surrounding the catheter at its tip over a distance of 0.5 cm. The response time (T 90) at 37°C is approximately 70 to 90 seconds. The diameter of the catheter is 0.5 mm. Internal zero drift of the catheter used is ≤0.6 mmHg and sensitivity drift <10% per 24 hrs.

Results

Positioning of the pO_2 Sensor for Ventricular CSF Monitoring

The pO_2 catheter was fixed in a stereotactic drive. After penetration of the dura the subarachnoid CSF

Table 1. *CSFpO$_2$ in Subarachnoid Space, Lateral Ventricle and Cisterna Magna under Normocapnic Conditions at Slightly Elevated Arterial pO$_2$ Value*

	Ventricular CSFpO$_2$ (mmHg) ± SD	Cisternal CSFpO$_2$ (mmHg) ± SD	Brain tissue pO$_2$ (mmHg) ± SD	Subarachnoid CSFpO$_2$ (mmHg) ± SD	Arterial paO$_2$ (mmHg) ± SD
Cats (n = 6)	70.8	65.8		41.2	124.7
	6.5	9.3		4.7	8.0
Dogs (n = 5)	73.5			47.2	118.9
	4.5			4.9	7.7
Dogs (n = 7)	65.3		28.2		112.2
	12.6		±7.5		6.3

pO_2 was measured for five minutes. The average value in cats was 41.2 ± 4.7 and in dogs 47.2 ± 4.9 (Table 1). The sensor was then slowly driven deeper; during passage of the tissue, variable values of 15 to 30 mmHg were measured. Upon reaching the superior ventricular wall the pO_2 increased suddenly to a level of 60 to 80 mmHg. In contrast to measuring in tissue, within CSF the probe signal was not changed if the probe was moved. For measurement of ventricular CSF pO_2 the probe tip was positioned approximately midway between the superior and inferior ventricular walls. A probe position was accepted to be intraventricular only if the probe signal was stable during the pre-experimental observation period of one hour (at unchanged reference conditions), and only if the signal was not altered by upward and downward probe displacements of two mm each way. Probes in a definite intraventricular position were fixed with Histoacryl to the skull, thus avoiding CSF leakage and probe displacement.

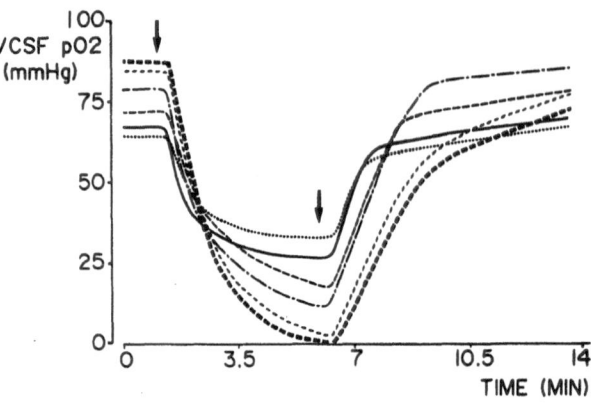

Fig. 1. Effect of decreasing FIO_2 on CSF pO_2. FIO_2: 16%, 14%, 12%, 10%, 8%, 6%

Reference Conditions

The average values ($\pm$SD) of the pO_2 obtained under reference conditions are summarized in Table 1; the average ventricular CSF pO_2 of dog and cat lay close to each other (73.5 ± 4.5 mmHg in dogs, and 70.8 ± 6.5 mmHg in cats). These values were obtained at slightly raised arterial pO_2 (124.7 ± 8.0 in cats and 118.9 ± 7.7 in dogs).

Ventilatory Experiments: Hypoxia, Hyperoxia, Hypocapnia and Hypercapnia

Changes of ventilatory conditions led to significant and fast changes in CSF pO_2. The results are summarized in Table 2. Changes in CSF pO_2 were observed 17 ± 2 seconds and maximal changes 31 ± 5 seconds after decrease of FIO_2. The steepest slope of the CSF pO_2 was observed after cessation of hypoxia. The post-hypoxic V CSF pO_2 was higher than the pre-hypoxic pO_2. The effect of different levels of hypoxia on CSF pO_2 in a single experiment is illustrated in Fig. 1.

At hyperoxia increasing arterial pO_2 from 125 ± 8 mmHg to 282 ± 22 mmHg elicited an increase in V CSF pO_2 and C CSF pO_2 of 20% and 31% respectively. At higher arterial pO_2 levels the CSF pO_2 showed only small further increases; it therefore appears that CSF pO_2 is regulated (Fig. 2). Hyperventilation in the cat experiments decreased the V CSF pO_2 by 29% and cisternal CSF pO_2 by 28% (Fig. 3). Hypoventilation resulted in an average arterial pCO_2 of 53 ± 3 mmHg. Ventricular CSF pO_2 decreased only slightly, probably due to a concomitant decrease of arterial pO_2 from $125 \pm$

Table 2. *CSFpO₂ during Ventilatory Experiments*

	Ventricular CSFpO$_2$ mmHg $\pm$ SD	Cisternal CSFpO$_2$ mmHg $\pm$ SD	Arterial pO$_2$ mmHg $\pm$ SD	Arterial pCO$_2$ mmHg $\pm$ SD
Reference conditions	70.8	65.8	124.7	37.8
	6.5	9.3	8.0	3.1
Hypoxia (FiO$_2$ 0.15)	44.2	46.3	62.3	35.4
	3.2	6.1	6.6	1.6
Hyperoxia (FiO$_2$ 0.5)	87.3	95.4	281.6	37.4
	9.4	17.1	22.0	1.2
Hyperventilation	50.2	47.3	137.5	21.3
	4.1	10.4	17.0	1.3
Hypoventilation	66.4	74.4	91.4	53.4
	11.1	9.2	9.1	3.0
Hypercarbia	86.3	95.6	121.7	63.1
	12.3	10.2	6.3	5.4

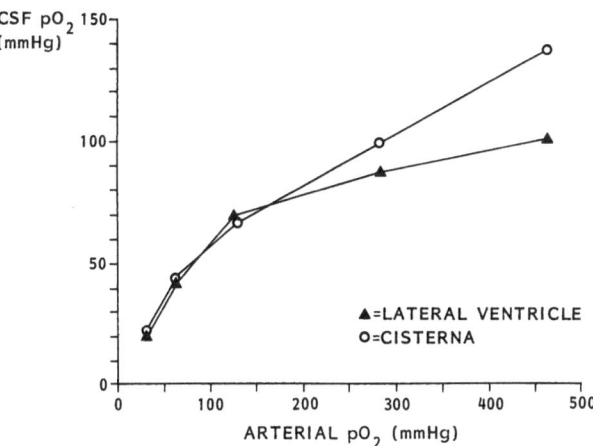

Fig. 2. Evidence for autoregulation of CSF pO$_2$

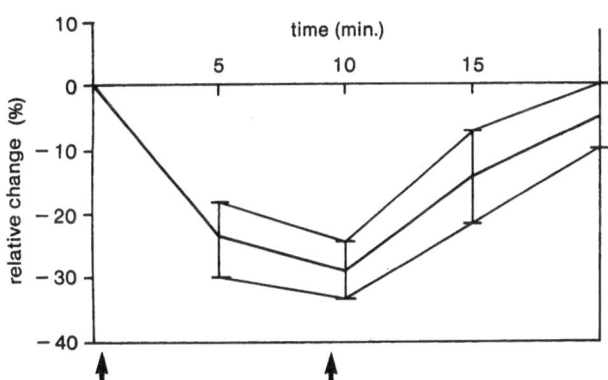

Fig. 3. Effect of hyperventilation on CSF pO$_2$

8 mmHg to 91 ± 9 mmHg. In contrast to the hypoventilation experiments hypercarbic ventilation (5% CO$_2$ admixture) increased both V CSF pO$_2$ and C CSF pO$_2$ values significantly.

Compared with the alterations of V CSF pO$_2$ in cats during hypoventilation, hyperventilation and hypercarbia, similar results were obtained in dog experiments.

Increased Intracranial Pressure and Decreased Cerebral Perfusion Pressure

In the experiments in which ICP was raised without preventing a secondary increase of blood pressure i.e. in which CPP did not decrease, no effect was seen on brain tissue pO$_2$ or CSF pO$_2$ (Table 3). In the experiments in which blood pressure was controlled by pericardiac tamponade, and thus a compensatory increase of CPP was effectively prevented, a decrease of both brain tissue pO$_2$ and CSF pO$_2$ was noted (Table 4, Fig. 4). Although there was a slight decrease in brain tissue pO$_2$ when reducing CPP from 120 to 100 mmHg, a sharper decrease was noted below a cerebral perfusion pressure of 80 mmHg. Interestingly enough the brain temperature, measured simultaneously with brain tissue pO$_2$ decreased suddenly when CPP fell below 40 mmHg. In some experiments CPP was reduced to values of 0 or below (ICP ≥ mean arterial blood pressure). Brain tissue pO$_2$ in the absence of cerebral perfusion remained at 7 ± 1.6 mmHg and CSF pO$_2$ at 1.4 ± 1 mmHg. During reperfusion after decreasing ICP again, a discrepancy between brain tissue and CSF pO$_2$ was observed: CSF pO$_2$ was rapidly restored to normal or even above normal levels, brain tissue pO$_2$, however, remained low or even decreased further (Fig. 5).

Table 3. *Brain Tissue pO$_2$ and CSF-pO$_2$ at Increased ICP, but Normal CPP*

	ICP	CPP	MABP	PaO$_2$	PaCO$_2$	Br.T. pO$_2$	VCSF pO$_2$
Without BP control	11.3 ± 3.8	102 ± 22	113 ± 25	112 ± 6	39 ± 5	28 ± 7	65 ± 12
	38 ± 3	108 ± 21	122 ± 17	114 ± 8	35 ± 5	28 ± 9	63 ± 15
	57 ± 3	110 ± 19	168 ± 18	112 ± 13	38 ± 5	29 ± 10	59 ± 13

CPP	Cerebral perfusion pressure.
MABP	Mean arterial blood pressure.
Br.T. pO$_2$	Brain tissue pO$_2$.
VCSF pO$_2$	Ventricular CSF pO$_2$.

Table 4. *Brain Tissue pO$_2$ and CSF-pO$_2$ at Decreased CPP*

	ICP	CPP	MABP	PaO$_2$	PaCO$_2$	Br.T. pO$_2$	VCSF pO$_2$
With BP control	12 ± 5	122 ± 10	134 ± 8	111 ± 11	39 ± 7	31 ± 11	60 ± 7
	39 ± 2	61 ± 2	100 ± 2	105 ± 6	40 ± 8	22 ± 8	45 ± 9
	51 ± 4	51 ± 2	102 ± 3	106 ± 8	41 ± 5	19 ± 8	39 ± 10
	59 ± 1	40 ± 2	99 ± 2	104 ± 7	43 ± 4	15 ± 7	34 ± 10
	65 ± 5	29 ± 2	94 ± 4	102 ± 11	45 ± 8	12 ± 6	25 ± 13

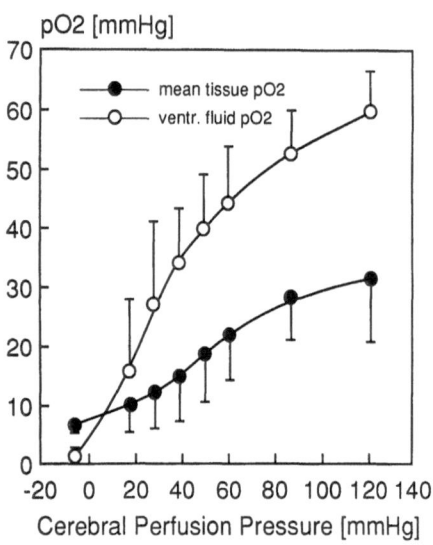

Fig. 4. Effect of decreasing CPP on CSF and brain tissue pO$_2$

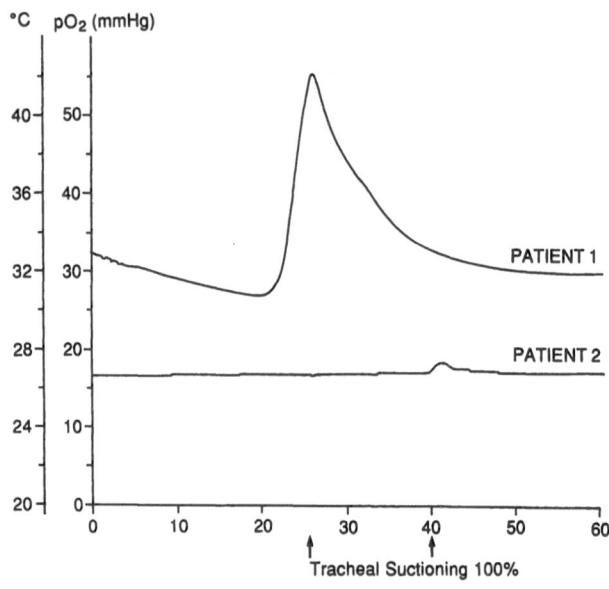

Fig. 6. Variable response of brain tissue pO$_2$ during hyperventilation with 100% oxygen

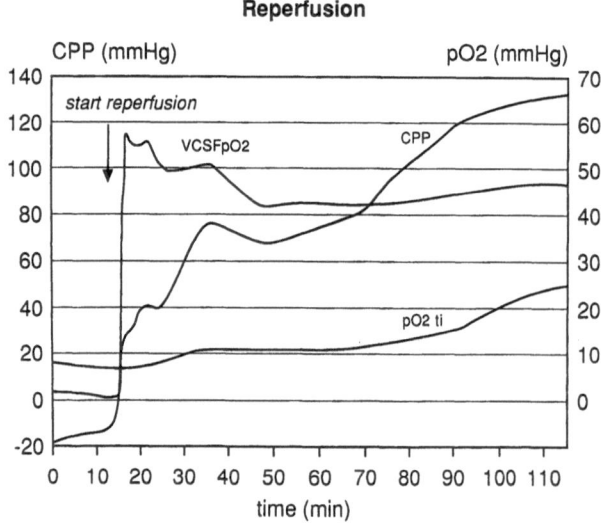

Fig. 5. Dissociation of brain tissue and CSF pO$_2$ after complete ischaemia

Brain Tissue pO$_2$ in Patients with Severe Head Injury

Preliminary measurements of brain tissue pO$_2$ have been performed in six patients with severe head injury. Stable measurements were obtained in these patients over various days of monitoring. Normal values of tissue pO$_2$ in the brain were similar to those obtained in animal experiments (25–30 mmHg). Changes in brain tissue pO$_2$ were observed when changing ventilation settings, and especially during manual hyperventilation with 100% oxygen during tracheal suctioning procedures (Fig. 6). Remarkably great differences in response of brain tissue pO$_2$ to short periods of ventilation with 100% oxygen were noted between individual patients. In one patient brain tissue pO$_2$ was monitored during the development of tentorial herniation and cerebral death (Fig. 7). Brain tissue pO$_2$ dropped

Brain Tissue pO₂ During Development of Cerebral Death

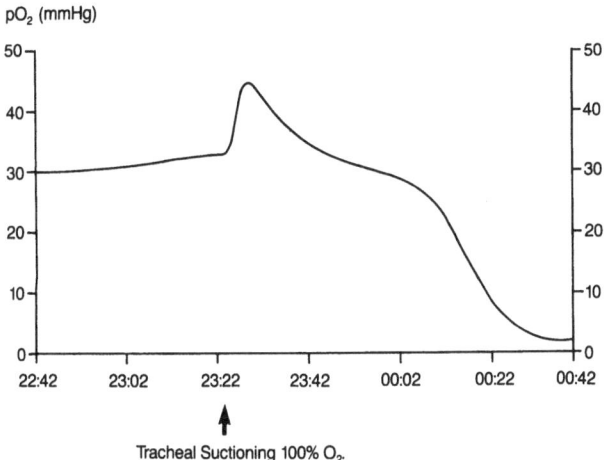

Fig. 7. Brain tissue pO_2 during development of cerebral death in a patient with severe head injury

to a level of about 5 mmHg. In this patient the decline in brain tissue pO_2 preceeded the clinical manifestation of cerebral herniation leading to brain death.

Discussion

CSF oxygen tension in the lateral ventricle of artificially ventilated cats and dogs were approximately the same. The results reported in this paper were obtained with a slight arterial hyperoxia. Calculating the ventricular CSF pO_2 at an arterial pO_2 of 100 mmHg from the regression of arterial pO_2 against ventricular CSF pO_2, results in CSF pO_2 values of 61 to 64 mmHg. Ventricular CSF pO_2 values in the same range have been reported in pigs as well as in man[5]. The reproducibility and stability of CSF pO_2 in different species – even after more than three hours of experimentation – suggest that CSF pO_2 is regulated. Furthermore, the lack of a further increase in CSF pO_2 at increasing arterial hyperoxia strongly suggests an autoregulatory mechanism for CSF pO_2.

In previous studies on CSF pO_2, samples from lumbar, cisternal or ventricular CSF were examined in blood gas analysors. This method, however, is considered unsuitable for O_2 measurements in liquids of low oxygen capacity, since oxygen dissolves and diffuses through the plastic material of tubes and chambers. Using such devices we obtained pO_2 readings up to 22 mmHg if physiological saline saturated with N_2 was analysed. Similar problems arise if hyperoxic samples of low oxygen capacity are analysed. In the 1960's already polarographic needle probes were used for continuous measurement of CSF pO_2. In most of these studies, however, the sensitivity of the probes drifted considerably, and this is possibly one of the reasons why pO_2 monitoring has not yet gained so much attention. The fast pO_2 probes used in our studies on CSF pO_2 drifted less than 2% per hour; recently the sensor has been further perfected and its drift is considerably less. Moreover, as the probe drift has been demonstrated to be linear, correction can be performed by recalibration calculations. It remains important to correct measured pO_2 values for brain temperature. For brain tissue pO_2 measurements a slower reacting pO_2 microsensor was used; the drift of the currently manufactured sensors in <5%/24 h, so that this sensor appears suitable for clinical measurements over some days.

Our studies show that CSF pO_2 and brain tissue pO_2 are influenced by changes in arterial pO_2, arterial pCO_2 and cerebral perfusion pressure. The observed fast pO_2 changes in response to changing ventilatory conditions are not easily explained if diffusion is the only oxygen transfer within cerebrospinal fluid. Neither can the time course of pO_2 changes be explained by bulk flow of CSF. An additional convectional oxygen transport in CSF must be assumed. The dynamics of oxygen transfer in CSF have previously been extensively discussed[6].

The CSF pO_2 changes consequent upon variations in arterial pCO_2 were much slower than those induced by variations in paO_2. It is inferred that changes in CSF pO_2 secondary to changes in arterial pO_2 directly reflect changes in oxygen offer, while changes in CSF pO_2 caused by arterial pCO_2 changes are mediated by differences in oxygen offer secondary to changes in cerebral blood flow. The post ischaemic dissociation between CSF pO_2 and brain tissue pO_2 as observed after very low cerebral perfusion pressures probably reflects the absence of oxygen demand of brain tissue.

It is concluded that both CSF pO_2 and brain tissue pO_2 reflect changes in cerebral oxygenation caused by changes in oxygen offer as well as by changes in cerebral blood flow; brain tissue pO_2 is also sensitive to the oxygen demand of cerebral tissue.

Recent studies on cerebral oxygenation have focussed on continuous monitoring of oxygen saturation in cerebral venous blood by means of jugular oxymetry[2, 3, 4, 9, 10, 12]. An oxygen saturation of 50% in the jugular bulb is considered critical, lower values indicating development of cerebral ischaemia. High values of oxygen saturation in the jugular bulb indicate hyperperfusion. The question remains however whether even at high oxygen saturation levels in the jugular bulb ischaemia may exist in the brain tissue itself. During hyperaemia blood can be preferentially shunted through short capillaries, with less effective oxygen transfer to the tissue than during normal blood flow conditions through long cerebral capillaries. Furthermore, the technique of jugular oxymetry is prone to technical problems. Problems concern position of the catheter tip resulting, in low light intensity and inaccurate measurements. These problems have been shown to occur in at least 30% of the measurements performed (Dearden, personal communication). Therefore, from both the theoretical and also the practical point of view direct monitoring of pO_2 in brain tissue or CSF as parameter for cerebral oxygenation would seem more appropriate. The results of our experimental studies indeed show these parameters to be indicative of cerebral oxygenation. For clinical use we preferred to measure tissue pO_2. First because this technique can also be used in the absence of a ventricular catheter, and secondly because of the observed dissociation between CSF and brain tissue pO_2 in the reperfusion phase after complete ischaemia. The results of the studies on the relation between cerebral perfusion pressure and brain tissue or CSF pO_2 indicate that a CPP of approximately 80 mmHg appears to be critical. Below this level brain tissue and CSF pO_2 decrease by 3 and 6 mmHg respectively per 10 mm decrease in CPP. In the CPP range of 40 to 60 mmHg a sharper decline is noted, indicating developing ischaemia. Chan *et al.*[1] have reported a critical CPP level of 70 mmHg in patients with severe head injury. Our results would support this conclusion that critical levels of CPP may well be higher than previously reported. Moreover, it may be that the critical level of CPP is different in individual patients. Monitoring brain tissue pO_2 could potentially ascertain the critical CPP level in individual patients. It is our opinion that monitoring brain tissue pO_2 could be of value as a warning against impending ischaemia, especially in patients with raised intracranial pressure, treated with hyperventilation. A disadvantage of the technique used is that only local brain tissue pO_2 is measured.

Changes of brain tissue pO_2 may of course occur in other regions of the brain not monitored. The primary object in monitoring brain tissue pO_2 however would at this stage of development of the technique not be to measure in areas of localized contusional damage, but more to monitor cerebral oxygenation in regions which are still relatively undamaged. For clinical measurements a modification of the subarachnoid screw was developed allowing for passage of three catheters into the brain tissue through one small burr hole. With this newly developed three way screw a fiberoptic pressure transducer can be positioned together with the pO_2 sensor and thermocouple probe. Further studies are in progress to validate the technique and establish the relative value of brain tissue pO_2 monitoring with respect to other techniques used for monitoring cerebral oxygenation, such as jugular oxymetry.

References

1. Chan KH, Dearden NM, Miller JD (1993) Multimodality monitoring of intracranial pressure therapy after severe brain injury. In: Avezaat CJJ, *et al* (eds) Intracranial pressure, Vol 8. Springer, Berlin Heidelberg New York Tokyo (in press)
2. Cold GE (1989) Does acute hyperventilation provoke cerebral oligaemia in comatose patients after acute head injury? Acta Neurochir (Wien) 96: 100–106
3. Cruz J, Minor ME, Allen SJ, Alves WM, Gennarelli TA (1993) Relationship between cerebral oxygenation and perfusion pressure in acute brain injury. In: Avezaat CJJ, *et al* (eds) Intracranial pressure, Vol 8. Springer, Berlin Heidelberg New York Tokyo (in press)
4. Dearden NM (1991) Jugular bulb venous oxygen saturation monitoring in the management of severe head injury. Curr Opin Anaesthesiol 4: 279–286
5. Fleckenstein W, Nowak G, Kehler U, Maas AIR, Dellbrügge HJ, De Jong DA, Hess M, Nollert G (1990) Oxygen pressure measurements in cerebrospinal fluid. Med Tech 110: 44–53
6. Fleckenstein W, Maas AIR, Nollert G, De Jong DA (1990) Oxygen pressure in Cerebrospinal Fluid. In: Ehrly AM, *et al* (eds) Clinical oxygen pressure measurement, Vol 2. Blackwell Ueberreuter Wissenschaft, Berlin, pp 368–395
7. Graham DI, Adams JH, Doyle D (1978) Ischaemic brain damage in fatal non-missile head injuries. J Neurol Sci 39: 231–234
8. Mendelow AD, Allcutt DA, Chambers I, Jenkins A, Crawford PJ, Sultan H (1993) Intracranial and cerebral perfusion pressure monitoring in the head injured patient: which index? In: Avezaat CJJ, *et al* (eds) Intracranial pressure, Vol 8. Springer, Berlin Heidelberg New York Tokyo (in press)
9. Obrist WD, Langfitt ThW, Jaggi JL, Cruz J, Gennarelli ThA (1984) Cerebral blood flow and metabolism in comatose patients with acute head injury. Relationship to intracranial hypertension. J Neurosurg 61: 241–253
10. Robertson CS, Narayan RK, Gokaslan ZL, Pahwa R, Grossman RG, Caram P, Allen E (1989) Cerebral arteriovenous oxygen difference as an estimate of cerebral blood flow in comatose patients. J Neurosurg 70: 222–230

11. Rosner MJ, Rosner SD (1993) Cerebral perfusion pressure management of head injury. In: Avezaat CJJ, *et al* (eds) Intracranial pressure, Vol 8. Springer, Berlin Heidelberg New York Tokyo (in press)
12. Sheinberg M, Kanter MJ, Robertson CS, Contant CF, Narayan RK, Grossman RG (1992) Continuous monitoring of jugular venous oxygen saturation in head injured patients. J Neurosurg 76: 212–217

Correspondence and Reprints: Dr. A. I. R. Maas, Department of Neurosurgery, Academic Hospital Rotterdam-Dijkzigt, 3015 GD Rotterdam, The Netherlands.

Acta Neurochir (1993) [Suppl] 59: 58–63

Studies of Tissue PO$_2$ in Normal and Pathological Human Brain Cortex

J. Meixensberger[1], **J. Dings**[1], **H. Kuhnigk**[2], and **K. Roosen**[1]

Departments of [1] Neurosurgery and [2] Anaesthesiology, University of Würzburg, Würzburg, Federal Republic of Germany

Summary

Brain cortex PO$_2$ was measured after craniotomy and opening of the dura mater in 26 patients. We determined the brain tissue PO$_2$ under standard narcotic conditions and after changing arterial PO$_2$ and PCO$_2$.

Patients were divided into two groups (normal and pathological), depending on the aspect of their cortex on Ct/MRI and intraoperative appearance of the cortex. No statistical significantly difference was seen between tissue PO$_2$ of the normal and the pathological group. A significant difference was seen only between the normal group and a subgroup with brain swelling (p = 0.0344). In the normal group no correlation was seen between tissue PO$_2$ and arterial PO$_2$ (r = 0.1541, p = 0.3076), whereas in the pathological group and especially in the oedema subgroup there was a highly significant correlation between tissue PO$_2$ and PaO$_2$ (r = 0.754, p = 0.0015 and r = 0.888, p = 0.0007).

Breathing 100% oxygen changed tissue PO$_2$ to 137.8 or 352 mmHg in the normal or the pathological group, respectively. Again, there was no correlation between tissue PO$_2$ and PaO$_2$ in the normal group (r = 0.1071, p = 0.392), whereas this correlation was significant in the pathological and the oedema subgroup (r = 0.6291, p = 0.0473 and r = 0.8385, p = 0.0185). This is evidence for regulatory mechanisms of tissue PO$_2$. During hyperventilation no significant difference in tissue PO$_2$ between the normal and the pathological group was seen. Low tissue PO$_2$ values, however, indicate a risk for inducing ischaemia.

Keywords: Tissue PO$_2$; brain cortex; brain oedema; O$_2$-metabolism.

Introduction

Haemodynamic changes and alterations of cerebral metabolism following brain damage have been the subject of numerous experimental and clinical investigations[4, 8, 12, 15–19, 21–26].

Although the understanding of cerebral haemodynamics and metabolism has immensely increased during the last years, there are still many questions to be answered. As the cerebral metabolism mainly depends on aerobic glycolysis, determination of oxygen pressure might give closer insight into the metabolic status or metabolic changes, as well as into cerebral haemodynamics. Thus monitoring of brain cortex PO$_2$ was started to clarify whether

- there are diffences in tissue PO$_2$ of normal and pathological human brain cortex and
- tissue PO$_2$ is influenced by changes of arterial PO$_2$ and PCO$_2$.

Material and Methods

26 patients (15 female, 11 male) were studied, whose mean age was 49.9 years (range 22–78). Table 1 lists the patients investigated, giving age, sex, diagnosis and location. Furthermore it was noted whether the cortex was to be considered normal or pathological, the extent of oedema on Ct/MRI and if there was brain swelling after opening the dura mater. No concomitant cardiovascular or pulmonary disease was present. Moreover, transcranial dopplersonography was performed preoperatively. Mean flow velocity was measured in the middle cerebral artery on both sides at different PaCO$_2$ whenever possible.

The brain tissue PO$_2$ measurements were performed after craniotomy and opening of the dura mater. All patients were submitted to standardized anaesthetic conditions, using dormicum, fentanyl, pancuronium and sometimes isoflurane for sedation, analgesia and relaxation. The gas mixture of O$_2$/N$_2$ contained a mean volume % of 39.4 O$_2$. No halothan was used, because of interference with the sensor qualities. The additional use of barbiturates was avoided.

The sensor used, is a surface sensor, coupled to a licox oxygen pressure measuring device, manufactured by G.M.S. Kiel-Mielkendorf, F.R.G. The sensor consists of 8 platinum cathodes and a silver anode. The oxygen that can freely diffuse through the sensor membrane and the electrolyte solution is reduced at the cathode to hydroxylions, changing the polarisation current. Current changes are proportional to the PO$_2$. The reaction time is <20 seconds. The measured PO$_2$ value is the mean of 8 single values of the 8 cathodes. The temperature coefficient of the sensor sensitivity is 2.4–2.5%/°c. Since a temperature probe is integrated in the sensor, the PO$_2$ values are automatically corrected when the tissue temperature changes. The sensor has a diameter of 15 mm and is 11 mm high. The measuring

Table 1. *Clinical Data of Patients Investigated*

No.	Initials	Age	Sex	Diagnosis	Location	Cortex*	Oedema**	Swelling***
1	F.E.	48	F	meningioma	front.	n	a	o
2	G.R.	25	M	ependymoma	pariet.	n	a	o
3	H.M.	23	F	ependymoma	temp.	n	a	o
4	H.A.	65	F	meningioma	pariet.	n	a	o
5	R.R.	52	F	meningioma	front.	n	a	o
6	S.R.	55	F	meningioma	front.	n	a	o
7	S.T.	22	F	cavernoma	temp.	n	a	o
8	W.B.	44	F	meningioma	sella	n	a	o
9	S.C.	35	F	aneurysm	ica	n	a	o
10	T.D.	49	M	cavernom	temp.	n	a	o
11	C.G.	48	F	aneurysm	multiple	n	a	o
12	C.G.	48	F	aneurysm	multiple	n	a	o
13	S.J.	28	M	oligo-astrocytoma (who3)	front.	n	a	o
14	G.K.	45	M	metastasis	front.	p	d	++
15	G.W.	41	M	meningioma	front.	p	d	o
16	I.G.	78	M	metastasis	front.	p	d	+
17	K.G.	61	M	metastasis	temp.	p	c	o
18	M.C.	60	M	glioblastoma	occip.	p	c	++
19	M.C.	45	M	metastasis	pariet.	p	a	++
20	P.E.	62	F	glioblastoma	temp-occip.	p	d	++
21	R.R.	57	M	glioblastoma	temp.	p	d	+
22	T.P.	61	F	glioblastoma	temp-pariet.	p	d	++
23	H.E.	64	F	meningioma	temp.	p	d	++
24	Z.M.	63	F	meningioma	front.	p	a	++
25	S.W.	57	M	aneurysm	acoa	p	a	o
26	G.T.	62	F	aneurysm	mca	p	a	o

  * Brain cortex n = normal, by Ct-MRI, intraoperative view and no brain swelling, p = pathological.
 ** Edema on Ct/MRI: a = no oedema, b = minimal, c = moderate, d = much.
*** Swelling after opening the dura mater: o = none, + = little, ++ = much.

surface area has a diameter of 5 mm. Before starting a measurement, a 2-point calibration (one calibration at an oxygen pressure of 0 mmHg and one at normal air oxygen pressure) has to be performed which takes about 45 min.

After opening of the dura mater, the sensor was placed on the cortex. Whenever possible, it was tried to avoid a placement of the sensor -on pial vessels. No CSF or blood was present between the cortex and the sensor. First basic tissue PO$_2$ was measured under steady state conditions in different areas. Then, the inspired oxygen fraction was increased up to 100%. After reaching a plateau in tissue PO$_2$, the inspired oxygen fraction was set back to the initial value. When tissue PO$_2$ had reached the stable initial value, the PaCO$_2$ was decreased by hyperventilation. After confirmation of a decreased PaCO$_2$ by blood gas analysis, the initial parameters were reinstalled. Before, during and after these tests, arterial blood gas analyses were made for determination of pH, BE, PO$_2$ and PCO$_2$. In addition MABP, rectal temperature and endtidal CO$_2$ were recorded. Tissue PO$_2$ and tissue temperature were automatically recorded every 5 seconds by a connected notebook.

Patients were assigned to the normal group when Ct/MRI showed no oedema, the operative aspect of the cortex was normal and there was no brain swelling after opening the dura mater. Patients who did not fulfil these criteria were assigned to the pathological group which was split into two subgroups, one with oedema on Ct/MRI and one subgroup with brain swelling after opening of the dura mater.

Results

Preoperative transcranial dopplersonography revealed normal blood flow velocities and all patients showed an intact CO$_2$-reactivity in both middle cerebral arteries.

Table 2 gives the basic tissue PO$_2$ data of the normal and pathological group, as well as of the two subgroups of the latter. No significant differences in tissue PO$_2$ between the normal and pathological group were noted (47.9 vs. 41.7 mmHg). However, there was a far wider range of tissue PO$_2$ in the pathological group (13–116 vs. 31–71 mmHg). Tissue temperature as well as PaO$_2$ were significantly different, though p = 0.0253 or p = 0.0347 respectively. The significant difference in PaO$_2$ between the normal and the pathological group is considered to be incidental.

A statistically significant difference in tissue PO$_2$ was seen between the normal group and the subgroup with brain swelling (47.9 vs. 33.71 mmHg, p

Table 2. *Basic Tissue PO$_2$ and Temperature in Normal and Pathological Brain Cortex as well as PaO$_2$, PaCO$_2$, pH, BE, MABP and Rectal Temperature under Narcotic Conditions*

	Normal (n = 13)	Pathological (n = 13)	Brain swelling (n = 7)	Oedema (n = 9)
Tissue PO$_2$	47.9 (13.14)	41.7 (30.83)	33.71 (19.62)	50.4 (33.66)
Tissue temp.	33.84 (1.01)	32.49 (2.08)	32.97 (0.76)	31.81 (2.08)
PaO$_2$	170.3 (28.1)	146 (36.5)	144.14 (35.6)	146.4 (39.8)
PaCO$_2$	30.01 (3.1)	30.42 (2.7)	29.91 (3.5)	30.08 (1.9)
pH	7.506 (0.04)	7.523 (0.04)	7.533 (0.04)	7.533 (0.04)
BE	2.75 (1.7)	4.08 (3.1)	4.44 (3.4)	4.6 (2.7)
MABP	89.62 (9.4)	93.92 (8.8)	95.71 (10.3)	92.78 (8.2)
Rect. temp.	36.03 (0.73)	36.17 (0.93)	36.31 (0.91)	36.03 (0.88)

Values are given as mean (+/− SD) mm Hg or °C.

= 0.0344). Again, the difference in tissue temperature (33.84 vs. 32.97°C, p = 0.0306) as well as in PaO$_2$ (170.3 vs. 144.14 mmHg, p = 0.044) gained statistical significance. Figure 1 gives the correlation between the PaO$_2$ and the tissue PO$_2$ under standardized anaesthetic conditions, with normal ventilation. While there is no correlation in the normal group (r = 0.1541, p = 0.3076), there is a highly

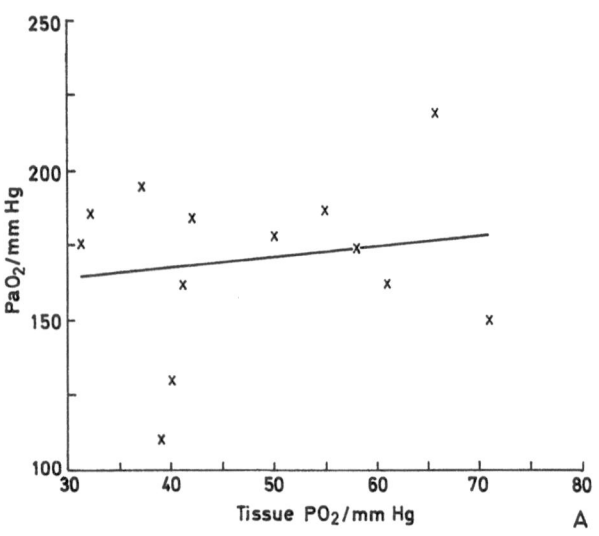

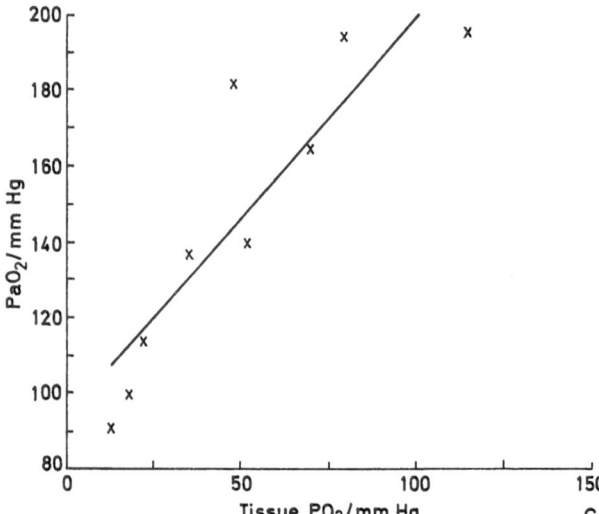

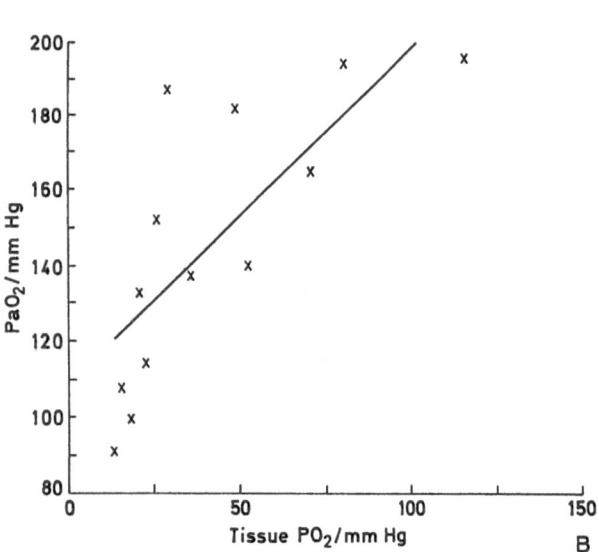

Fig. 1. (A) Correlation between PaO$_2$ and tissue PO$_2$ in the normal group: r = 0.1541, p = 0.3076, n = 13. (B) Correlation between PaO$_2$ and tissue PO$_2$ in the pathological group: r = 0.754, p = 0.0015, n = 13. (C) Correlation between PaO$_2$ and tissue PO$_2$ in the oedema subgroup: r = 0.888, p = 0.0007, n = 9

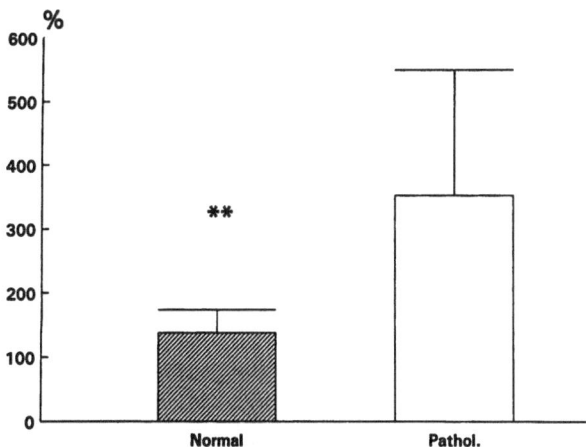

Fig. 2. % Change of brain cortex tissue PO$_2$ after breathing 100% oxygen in the normal (n = 9) and pathological (n = 8) group. Values are given as mean ($\pm$ SD)

statistical significantly correlation in the pathological group (r = 0.754, p = 0.0015), which is even stronger in the oedema subgroup (r = 0.888, p = 0.0007).

After changing the inhaled oxygen fraction to 1.0, tissue PO$_2$ increases to 107.9 mmHg in the normal group, and to 130.5 mmHg in the pathological group (difference n.s. p = 0.2261). The % change of tissue PO$_2$ of the normal and of the pathological group (137.8 vs. 352%) is significantly different (p = 0.0095) (Fig. 2). In the normal group no correlation between PaO$_2$ and tissue PO$_2$ was seen (r = 0.1071, p = 0.392). Again, there is a correlation in the pathological group (r = 0.6291, p = 0.0473), and

in the oedema subgroup (r = 0.8385, p = 0.0185).

Table 3 gives the data of all groups following hyperventilation. The only statistical significantly difference is seen in the tissue temperature of the normal and pathological group. The most pronounced changes in tissue PO$_2$ were seen in the normal group (14.5% vs. 7.7% in the pathological group). The lowest values of tissue PO$_2$ after hyperventilation were found in the pathological group (9.0 mmHg).

Discussion

Since the publication of Clark[3] in 1956 on the possibility of monitoring oxygen tension in blood and tissue, various experiments have been performed, to assess brain tissue PO$_2$ and its regulatory mechanisms[1, 2, 5–7, 9, 10, 13, 14, 20, 27–29]. So far there are only few reports on in vivo measurements of tissue PO$_2$ of human brain cortex[1, 10]. Assad et al.[1] used a polarographic multiwire surface electrode according to Lübbers and Kessler[11]. They studied brain tissue PO$_2$ in different brain tumours as well as in the adjacent oedematous and compressed cortex. At a PaO$_2$ of 150–170 mmHg and a PaCO$_2$ of 30 mmHg, in three patients they obtained average values of 33–36 mmHg in normal cortex. The difference in tissue PO$_2$ of the normal cortex between their values and ours (47.9 mmHg at PaO$_2$ 170.3 mmHg and PaCO$_2$ 30.01 mmHg) might be explaned by anaesthesia, (non) compensated tissue temperature or placement of the probe over pial

Table 3. *Tissue PO$_2$ and Temperature in Normal and Pathological Brain Cortex as well as PaO$_2$, PaCO$_2$, MABP and Rectal Temperature after Hyperventilation*

	Normal (n = 8)	Pathological (n = 9)	Brain swelling (n = 5)	Oedema (n = 8)
Tissue PO$_2$	43 (14.58)	45.33 (35.3)	32.2 (23.1)	48.5 (36.3)
% Change	14.5 (14.32)	7.7 (9.7)	10.2 (12.3)	7.3 (10.3)
Tissue temp.	34.8 (1.6)	32.5 (1.9)	33.1 (0.6)	32.4 (2.0)
% Change	1.26 (2.23)	2.5 (3.8)	1.3 (1.8)	2.8 (3.9)
PaO$_2$	171.3 (37.1)	146.1 (41)	133.4 (37)	145.5 (43.7)
% Change	6.1 (5.1)	8.8 (6.3)	9.2 (6.4)	8.2 (6.4)
PaCO$_2$	24.34 (2.0)	26.1 (2.3)	26.7 (3.0)	25.4 (0.7)
% Change	19.79 (7.83)	16.5 (5.0)	17.4 (6.4)	15.4 (3.8)
MABP	91.3 (12.9)	93.7 (8.4)	96.2 (8.9)	91.6 (6.2)
% Change	4.5 (3.8)	2.6 (3.5)	4.4 (4.0)	2.7 (3.8)
Rect. temp.	36.2 (0.9)	35.9 (0.7)	36.0 (0.8)	35.9 (0.8)
% Change	0.1 (0.15)	0.17 (0.16)	0.2 (0.16)	0.15 (0.16)

Values are given as mean (+/− SD) mm Hg or °C.

vessels. Assad's findings of lower values in compressed cortical areas and a wider range of values in oedematous cortex is in agreement with our findings. Kayama *et al.*[10] used a Clark surface sensor for measuring mean oxygen pressure in brain tissue surrounding various kinds of (metastatic) brain tumours. The mean tissue PO_2 was 59.8 ± 6.5 mmHg (n = 11) at a PaO_2 of 112.7 ± 5.2 mmHg. The difference with regard to our results [41.7 mmHg for the pathological and 50.4 mmHg for the edema (sub)group] may be caused by the above mentioned factors.

Data of human brain tissue PO_2 during hyperoxia and hyperventilation are not available. So far, only animal experiments have been done[7, 13, 20]. In these experiments various reactions to hyperoxia were seen: an increase, no change or a decrease of tissue PO_2. This lead to the postulation of a regulatory mechanism for local PO_2. Although all patients in this series who underwent hyperoxia showed an increase in brain tissue PO_2, there is still evidence for such a regulatory mechanism: In normal human brain tissue we found no correlation between PaO_2 and tissue PO_2, whereas in the pathological group there is a correlation between PaO_2 and tissue PO_2. This correlation persists during hyperoxia. As the hyperoxia lasted only until a plateau in tissue PO_2 was reached (after 3–5 min), there are no data about possible tissue PO_2 reactions during continuing hyperoxia. The effects of hyperventilation were as expected by other haemodynamic examinations[15]. The most pronounced changes of tissue PO_2 were seen in the normal group (14.5%). In this group, though, there were some patients with no reaction, indicating an impaired CO_2-reactivity. Simultaneous local cerebral blood flow measurements with laser doppler probes will probably give additional information about such disturbances. Finally, hyperventilation is capable of decreasing oxygen pressure to very low levels in the pathological group, indicating that some patients are at risk from ischaemia during hyperventilation.

Conclusions

Brain tissue PO_2 is altered in pathological human brain cortex. There is evidence for a regulatory mechanism of local tissue PO_2. Monitoring of brain tissue PO_2 may influence antiischaemic therapy.

References

1. Assad F, Schultheiss R, Leniger-Follert E, Wüllenweber R (1984) Measurement of local oxygen partial pressure (PO2) of the brain cortex in cases of brain tumors. Adv Neurosurg 12: 263–266
2. Bicher HI, Bruley D, Reneau DD, Knisely M (1973) Regulatory mechanisms of brain oxygen supply. In: Kessler M, *et al* (eds) Oxygen supply. Theoretical and practical aspects of oxygen supply and microcirculation of tissue. Urban and Schwarzenberg, München-Berlin-Wien, pp 180–185
3. Clark LC Jr (1956) Monitor and control of blood and tissue oxygen tensions. Trans Soc Art Int Organs 2: 41–48
4. Enevoldsen EM (1986) CBF in head injury. Acta Neurochir (Wien) [Suppl] 36: 133–136
5. Fleckenstein W, Maas AIR, Nollert G, de Jong DA (1990) Oxygen pressure in cerebrospinal fluid. In: Ehrly AM, *et al* (eds) Clinical oxygen pressure measurement II. Blackwell Ueberreuter Wissenschaft, Berlin, pp 368–395
6. Grote J, Zimmer K, Schubert R (1985) Tissue oxygenation in normal and edematous brain cortex during arterial hypocapnia. Adv Exp Med Biol 180: 179–184
7. Ingvar DH, Lübbers DW, Siesjö B (1960) Measurement of oxygen tension on the surface of the cerebral cortex of the cat during hyperoxia and hypoxia. Acta Physiol Scand 48: 373–381
8. Jaggi JL, Obrist WD, Gennarelli TA, Langfitt TW (1990) Relation-ship of early cerebral blood flow and metabolism to outcome in acute head injury. J Neurosurg 72: 176–182
9. Jamieson D, van den Brenk HAS (1963) Measurement of oxygen tensions in cerebral tissues of rats exposed to high pressures of oxygen. J Appl Physiol 18: 869–876
10. Kayama T, Yoshimoto T, Fujimoto S, Sakurai Y (1991) Intratumoral oxygen pressure in malignant brain tumor. J Neurosurg 74: 55–59
11. Kessler M, Lübbers DW (1966) Aufbau und Anwendungsmöglichkeiten verschiedener PO2-Elektroden. Pflügers Arch 291: R82
12. Langfitt TW, Obrist WD, Gennarelli TA, O'Connor MJ, Weeme AT (1977) Correlation of cerebral blood flow with outcome in head injured patients. Ann Surg 186: 411–414
13. Leniger-Follert E (1977) Direct determination of local oxygen consumption of the brain cortex in vivo. Adv Exp Med Biol 94: 325–330
14. Leniger-Follert E, Lübbers DW, Wrabetz W (1975) Regulation of local tissue PO2 of the brain cortex at different arterial O2 pressures. Pflügers Arch 359: 81–95
15. Meixensberger J, Brawanski A, Dannhauser-Leistner J, Holzschuh M, Ullrich W (1993) Is there a risk to induce ischemia be hyperventilation therapy? In: Avezaat CJJ, *et al* (eds) Intracranial pressure, Vol 8. Springer, Berlin Heidelberg New York Tokyo (in press)
16. Messeter K, Nordström C-H, Sundbärg G, Algotsson L, Ryding E (1986) Cerebral hemodynamics in patients with severe head trauma J Neurosurg 64: 231–237
17. Muizelaar JP, Lutz HA, Becker DP (1984) Effect of mannitol on ICP and CBF and correlation with pressure autoregulation in severely head-injured patients. J Neurosurg 61: 700–706
18. Muizelaar JP, Marmarou A, DeSalles AAF, Ward JD, Zimmerman RS, Li Z, Choi SC, Young HF (1989a) Cerebral blood flow and metabolism in severely head injured children. Part 1: Relationship with GCS score, outcome, ICP, and PVI. J Neurosurg 71: 63–71

19. Muizelaar JP, Ward JD, Marmarou A, Newlon PG, Wachi A (1989b) Cerebral blood flow and metabolism in severely head injured children. Part 2: Autoregulation. J Neurosurg 71: 72–76
20. Nair P, Whalen WJ, Buerk D (1975) PO2 of cat cerebral cortex: Response to breathing N2 and 100% O2. Microvasc Res 9: 158–165
21. Nilsson B, Nordström C-H (1977a) Experimental head injury in the rat. Part 3: Cerebral blood flow and oxygen consumption after concussive impact acceleration. J Neurosurg 47: 262–273
22. Nilsson B, Nordström C-H (1977b) Rate of cerebral energy consumption in concussive head injury in the rat. J Neurosurg 47: 274–281
23. Nordström C-H, Messeter K, Sundbärg G, Schalen W, Werner M, Ryding E (1988) Cerebral blood flow, vasoreactivity and oxygen consumption during barbiturate therapy in severe traumatic brain lesions. J Neurosurg 68: 424–431
24. Obrist WD, Langfitt TW, Jaggi JL, Cruz J, Gennarelli TA (1984) Cerebral blood flow and metabolism in comatose patients with acute head injury. Relationship to intracranial hypertension. J Neurosurg 61: 241–253
25. Robertson CS, Grossman RG, Goodman JC, Narayan RK (1987) The predictive value of cerebral anaerobic metabolism with cerebral infarction after head injury. J Neurosurg 67: 361–368
26. Robertson CS, Narayan RK, Gokaslan ZL, Pahwa R, Grossman RG, Caram P, Allen E (1989) Cerebral arteriovenous oxygen difference as an estimate of cerebral blood flow in comatose patients. J Neurosurg 70: 222–230
27. Silver JA (1973) Brain oxygen tension and cellular activity. In: Kessler M, et al (eds) Oxygen supply. Theoretical and practical aspects of oxygen supply and microcirculation of tissue. Urban and Schwarzenberg, München-Berlin-Wien, pp 186–188
28. Smith RH, Guilbeau EJ, Reneau DD (1977) The oxygen tension field within a discrete volume of cerebral cortex. Microvasc Res 13: 233–240
29. Whalen WJ, Ganfield R, Nair P (1970) Effects of breathing O2 or O2 + CO2 and of the injection of neurohumors on the PO2 of cat cerebral cortex. Stroke 1: 194–200

Correspondence and Reprints: Dr. J. Meixensberger, Neurochirurgische Klinik und Poliklinik der Universität Würzburg, Josef-Schneider-Strasse 11, D-97080 Würzburg, Federal Republic of Germany.

Acta Neurochir (1993) [Suppl] 59: 64–68

Laser-Doppler Flowmetry

Applicability of Laser-Doppler Flowmetry for Cerebral Blood Flow Monitoring in Neurological Intensive Care

R. L. Haberl, A. Villringer, and **U. Dirnagl**

Department of Neurology, Klinikum Grosshadern, University of Munich, Munich, Federal Republic of Germany

Summary

Laser Doppler flowmetry (LDF) is a technique for real-time assessment of cerebral blood flow (CBF) changes with potential clinical applicability. Experimental studies have validated that LDF allows accurate measurement of changes in CBF due to physiological and pathophysiological stimuli. Absolute quantitation of flow in ml/100 g min by LDF is not possible. The technique may be used in patients during open brain surgery and postoperatively for bedside CBF monitoring. Disadvantages of the technique are that the flow measurement is highly localized (about 1 mm^3) and artifacts may be produced by movement, light or probe placement over large surface vessels. The fibre optic probes for LDF are small enough to be introduced into routinely used intraventricular pressure catheters. We suggest that simultaneous monitoring of CBF and intracranial pressure by such a device holds promise for improved management of patients with critical brain injury.

Keywords: Laser-Doppler; cerebral blood flow; intensive care monitoring.

Introduction

There is a need for CBF monitoring in intensive care patients with severe brain disease. Currently, the maintenance of CBF is indirectly assessed by measurement of intracranial pressure (ICP) and cerebral perfusion pressure (CPP). However, the relationship of CPP and CBF is complex and the lower limit of CPP to be maintained to avoid clinically relevant cerebral ischaemia is a matter of controversy. The ultimate aim in surveilling those critically ill patients is continuous recording of cerebral blood flow and cerebral metabolic rate ($CMRO_2$) with high regional and temporal resolution.

There are several approaches to achieve this goal. The best technique to simultaneously measure CBF and $CMRO_2$ in patients is position emission tomography (PET). PET scanning, however, is not applicable as a monitoring technique because it cannot be used in the intensive care unit (ICU), has a low temporal resolution, and does not provide continuous recordings. The important, therapeutically relevant information is the immediate registration of a drop in CBF during acute medical intervention – like sedation – or nursing procedures – like posturing. There is only a limited number of techniques, which may provide this information and which may be used in patients in the ICU. Among those, laser Doppler flowmetry (LDF) may have advantages not shared by the other methods.

Methods

Principle of Laser Doppler Flowmetry

LDF is based on the direct detection of the velocity and number of red blood cells flowing through microvessels[3, 20]. It has the advantage that it provides continuous assessment of flow and the probes used to detect flow do not invade the tissue to be analyzed. Low-power laser light (energy at the probe tip: <1 mW) is directed to the tissue through an optical fibre. As light enters the tissue, photons are scattered in a random fashion by moving blood cells in microvessels and stationary tissue cells. Because of the random scattering events, the direction of blood flow is not a measurement parameter. Photons that interact with moving blood cells are Doppler (frequency) shifted and scattered. Photons that interact with stationary tissue cells are scattered but not Doppler shifted. A portion of the emitted light is reflected back into receiving fibers within the fibre-optic probe after multiple scattering events. The photons mix (heterodyne) on the surface of a photodetector and generate an electrical signal[3]. The fraction of the backscattered light that is Doppler shifted is processed to yield information on the velocity of flowing blood cells (proportional to the frequency shift) and on the blood volume (proportional to the power of frequeny shifted light) in arbitrary units. Blood flow is computed by determining the product of blood volume and blood velocity. The instrument measures these parameters in a tissue volume of 1 mm[3]. For theoretical reasons, the distance between the fibre-optic probe and the tissue surface, as well as motion artifacts (as produced by respiration) primarily affect the reading of volume (and thereby of flow), whereas the velocity reading is not sensitive to minor changes in probe position.

Results and Discussion

Validity of CBF Measurement by Laser-Doppler Flowmetry

LDF for CBF measurement was extensively validated by alternative techniques in experimental studies. In one of the first validating studies, Eyre et al. have shown good correlation of LD flow and microsphere measurement of CBF[9]. The study, however, may not be conclusive because microspheres measure flow in a tissue volume that is much greater than that measured by LDF and therefore does not represent flow in the same, highly localized microcirculation. Subsequent investigations using the hydrogen clearance technique[12, 13, 19] and [14C]iodo-antipyrine autoradiography[8] have shown excellent correlation of changes in LDF-CBF and other techniques examining highly localized CBF. In particular, alterations in corpuscular blood flow as assessed by LDF correspond to changes in CBF as measured by dilutional techniques like H_2-clearance or autoradiography. This argues against the possibility that transient "plasmatic" filling of brain capillaries – as seen during in vivo confocal microscopy of the rat brain cortex in about 4% of capillaries[21] – or moderate changes in microvascular haematocrit may flaw CBF measurements by LDF.

Importantly, LDF does not provide accurate measurements of *absolute* CBF values. Dirnagl et al. have shown no correlation between absolute CBF values as measured by autoradiography and arbitrary LDF units[8]. It is not justified, therefore, to introduce calibration factors to convert arbitrary LDF units into absolute flow readings (i.e., in ml/100 g min). While LDF is capable of quantitatively

recording on-line changes in CBF, the readings of absolute blood flow units with this method are meaningless.

Comparison of LDF with Alternative Techniques

Few techniques measure CBF continuously, are non-invasive and may therefore be used in patients for bedside monitoring of cerebral blood flow. Table 1 shows techniques with potential applicability on the ICU. All of them have been used experimentally in man, but only transcranial Doppler sonography has been developed into a routine procedure. There is no technique available which fulfills the "ultimate goal", i.e. simultaneous measurement of CBF and brain metabolism at the bedside. Validation studies are needed to compare the CBF information of those methods with values obtained by experimental standard techniques. This has only been done for LDF and for the thermodiffusion technique.

Clinical Applications of LDF

LDF has been used to measure CBF during brain surgery[2, 10, 17]. The authors concluded that intraoperative use might be an aid in surgery of arteriovenous malformations[17] and tumours[2]. In those studies an attempt was made to calibrate LDF for absolute quantitation of CBF, which is inappropriate as stated above. There is only one published observation on the application of LDF for postoperative blood flow monitoring in neurological intensive care patients[15]. Meyerson et al. described four comatose patients in whom LDF was simple to use and reliable. A 500 µm diameter probe was inserted subdurally, a second type of microprobe was used simultaneously for intracortical measurements at 2 mm below the surface. The probes were left in place for up to 13 days without complications. Notably, these authors found treatment induced situations where a minor reduction in systemic arterial blood pressure caused fluctuations in ICP and a marked decrease in CBF_{LDF}. Those episodes may be harmful for the patient. They remain unnoticed, when continuous CBF monitoring is lacking.

We have used LDF for continuous CBF monitoring in 8 patients on our neurological ICU. All of them were comatose because of severe head trauma (n = 2), life threatening bacterial meningitis (n = 3) or large brain infarction (n = 3). Ventricular catheters were inserted in all patients for intracranial

Table 1. *Techniques for Bedside CBF Monitoring*

Method	Evaluation of	Validated by	Spatial resolution	Advantage	Disadvantage	Literature
Laser Doppler flowmetry	microcirculatory CBF	hydrogen clearance, autoradiography	$1\,mm^3$	thouroughly validated, high temporal resolution	quantitation of relative changes only, craniectomy necessary	Stern 1975, Haberl *et al.* 1989, Dirnagl *et al.* 1989
Thermodiffusion	rCBF	^{133}xenon clearance, hydrogen clearance	$1-3\,cm^3$	quantitation of absolute CBF units, high temporal resolution	subdural probe position, craniectomy necessary	Carter *et al.* 1981, Gaines *et al.* 1983, Carter *et al.* 1991
Near infrared spectroscopy	cytochrome a/a$_3$ redox state, Hb/HbO$_2$, CBF and blood volume		several cm^3	genuinely non-invasive, high temporal resolution	low spatial resolution, CBF not validated	Jöbsis 1977, Villringer *et al.* 1993
Jugular venous oxygen saturation	global CBF		global brain	no craniectomy necessary	no spatial resolution, errors by probe displacement	Cruz 1986, Cruz *et al.* 1990, Sheinberg *et al.* 1992
Transcranial Doppler sonography	blood velocity in large cerebral arteries		major cerebral arteries	non-invasive	discontinuous, signal influenced by large artery caliber variation	Aaslid *et al.* 1982, Newell and Aaslid 1992

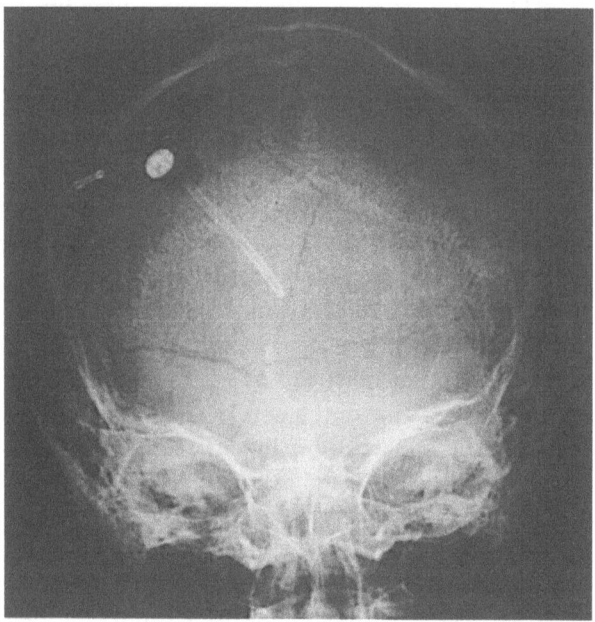

Fig. 1. Continuous bedside monitoring of CBF by LDF: a LD-disc probe is placed on the dura next to the ventricular pressure catheter. The catheter and the fibre optics of the Laser-Doppler are inserted through the same burrhole

pressure measurement. A LD-disc probe (8 mm in diameter) was inserted through the same burr hole and was placed on the dura next to the ventricular line (Fig. 1). The dura mater was left intact. In our animal studies with rats and rabbits, dampening of the laser signal by the dura was minimal. The absence of an extracerebral flow artifact from the dura, however, has still to be confirmed in humans. The fibre-optics, together with the ventricular catheter, were tunneled subcutaneously for 10 cm. A Vasamedics BPM 403a LD-flowmeter (Vasamedics, St. Paul, MN) was used. The time constant of the flowmeter was set at 1 second. The output LDF signal, systemic arterial pressure and ICP were recorded simultaneously on a strip chart recorder.

Figure 2 shows an example of the effect of sedation with midazolam on LD-flow, MABP and ICP. The marked decrease in flow with a minor drop in blood pressure stresses the potential hazards of routine therapeutic procedures. Controlled application of those agents and CBF adjusted posturing may be the merit of using LDF on the ICU.

We encountered a number of problems when using LDF as a bedside CBF monitor. A stable baseline reading could not be obtained in some patients, especially when they started to move spontaneously. Respiration artifacts were primarily seen in the volume- and the flow-reading, with velocity being quite stable. Dislocation of the probe and subsequent erratic readings occurred in two patients. These problems could be overcome by insertion

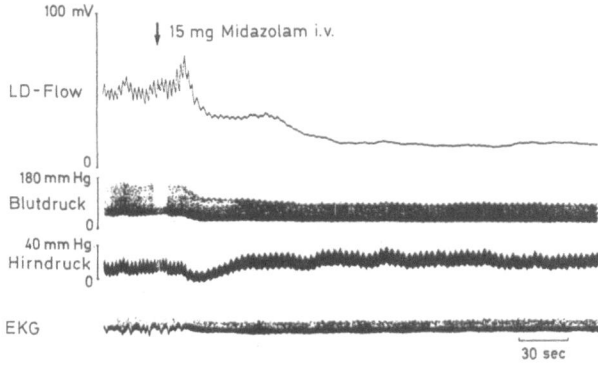

Fig. 2. Effect of sedation with 15 mg midazolam on CBF (*LD-flow*), arterial blood pressure (*Blutdruck*), intracranial pressure (*Hirndruck*) and ECG (*EKG*) in a patient with head trauma. Note the drastic drop in LD-flow with the sedation-associated decrease in arterial blood pressure and the moderate increase in ICP

of the LD-probe into the cortical tissue instead of placing it on the dura. Finally, it is difficult to remove the probes which we are currently using. The optical fibers were broken in one case when the probe was removed.

Perspectives

Some technical modifications are required to adapt LDF for reliable bedside CBF monitoring. Probe displacement, movement artifacts and erratic readings due to overlap with large surface vessels, may be minimized by positioning the probe below the surface into the cortical tissue. Meyerson *et al.* have shown that the surface LDF signal does not differ from that obtained by the intracortical probe[15]. This observation also argues against the possibility that tissue damage by the intracortical probe may affect LD flow. A device is currently developed in our laboratory where thin optical fibers for LDF are introduced into the wall of routinely used ventricular ICP catheters. We suggest that this may be a way to simultaneously measure ventricular pressure and CBF at several cortical and subcortical sites along the catheter.

Although LDF may be used as a simple technique to continuously assess cerebral blood flow changes at the bedside, its major limitation may be the small tissue volume of measurement. In particular, this becomes a problem when the probe is close to focal brain lesions where adjacent areas of hypoperfusion and hyperperfusion exist. One aim of our ongoing studies is to compare the CBF information from LDF, and possibly supplement it, with techniques, which evaluate CBF in larger tissue volumes like jugular venous oxygen saturation and near infrared spectroscopy.

References

1. Aaslid R, Markwalder TM, Nornes H (1982) Noninvasive transcranial Doppler ultrasound recording of flow velocity in basal cerebral arteries. J Neurosurg 57: 769–774
2. Arbit E, DiResta GR, Bedford RF, Shah NK, Galicich JH (1989) Intraoperative measurement of cerebral and tumor blood flow with laser-Doppler flowmetry. Neurosurgery 24: 166–170
3. Bonner RF, Clem TR, Bowen PD, Bowman RL (1981) Laser-Doppler continuous real-time monitor of pulsatile and mean blood flow in tissue microcirculation. In: Chen SH, Chu B, Nossal R (eds) Scattering techniques applied to supramolecular and nonequilibrium systems. Plenum Press, New York, pp 685–702
4. Carter LP, Ersparmer R, Bro WJ (1981) Cortical blood flow: thermal diffusion vs. isotope clearance. Stroke 12: 513–518
5. Carter LP, Grahm T, Bailes JE, Bichard W, Spetzler RF (1991) Continuous postoperative monitoring of cortical blood flow and intracranial pressure. Surg Neurol 35: 36–39
6. Cruz J, Miner ME (1986) Modulating cerebral oxygen delivery and extraction in acute traumatic coma. In: Miner ME, Wagner KA (eds) Neurotrauma treatment rehabilitation and related issues. Butterworths, Boston London Durban Singapore Sydney Toronto Wellington, pp 55–72
7. Cruz J, Miner ME, Allen SJ, Alves WM, Gennarelli TA (1990) Continuous monitoring of cerebral oxygenation in acute brain injury: injection of mannitol during hyperventilation. J Neurosurg 73: 725–730
8. Dirnagl U, Kaplan B, Jacewicz M, Pulsinelli W (1989) Continuous measurement of cerebral cortical blood flow by laser-Doppler flowmetry in a rat stroke model. J Cereb Blood Flow Metab 9: 589–596
9. Eyre JA, Essex TJ, Flecknell PA, Bartholomew PH, Sinclair JI (1988) A comparison of measurements of cerebral blood flow in the rabbit using laser Doppler spectroscopy and radionuclide labelled microspheres. Clin Phys Physiol Meas 9: 65–74
10. Fasano VA, Urciuoli R, Bolognese P, Mostert M (1988) Intraoperative use of laser Doppler in the study of cerebral microvascular circulation. Acta Neurochir (Wien) 95: 40–48
11. Gaines C, Carter P, Crowell RM (1983) Comparison of local blood flow determined by thermal and hydrogen clearance. Stroke 14: 66–69
12. Haberl RL, Heizer ML, Ellis EF (1989) Laser-Doppler assessment of brain microcirculation: effect of local alterations. Am J Physiol 256: H1255–H1260
13. Haberl RL, Heizer ML, Marmarou A, Ellis EF (1989) Laser-Doppler assessment of brain microcirculation: effect of systemic alterations. Am J Physiol 256: H1247–H1254
14. Jöbsis FF (1977) Noninvasive, infrared monitoring of cerebral and myocardial oxygen sufficiency and circulatory parameters. Science 198: 1264–1267
15. Meyerson BA, Gunasekera L, Linderoth B, Gazelius B (1991) Bedside monitoring of regional cortical blood flow in comatous patients using laser Doppler flowmetry. Neurosurgery 29: 750–755

16. Newell DW, Aaslid R (1992) Transcranial Doppler: clinical and experimental uses. Cerebrovasc Brain Metab Rev 4: 122–143

17. Rosenblum BR, Bonner RF, Oldfield EH (1987) Intraoperative measurement of cortical blood flow adjacent to cerebral AVM using laser Doppler velocimetry. J Neurosurg 66: 396–399

18. Sheinberg M, Kanter MJ, Robertson CS, Contant CF, Narayan RK, Grossman RG (1992) Continuous monitoring of jugular venous oxygen saturation in head-injured patients. J Neurosurg 76: 212–217

19. Skarphedinsson JO, Harding H, Thoren P (1988) Repeated measurements of cerebral blood flow in rats. Comparisons between the hydrogen clearance method and laser Doppler flowmetry. Acta Physiol Scand 134: 133–142

20. Stern MD (1975) In vivo evaluation of microcirculation by coherent light scattering. Science 254: 56–58

21. Villringer A, Dirnagl U, Gebhardt R, Einhäupl KM (1991) An in vivo approach to assess the capillary recruitment hypothesis in the brain microcirculation using confocal laser scanning microscopy. J Cereb Blood Flow Metab 11 [Suppl 2]: 441

22. Villringer A, Planck J, Hock C, Schleinkofer L, Dirnagl U (1993) Near infrared spectroscopy (NIRS): a new tool to study hemodynamic changes during activation of brain function in human adults. Neurosci Lett 154: 101–104

Correspondence and Reprints: R.L. Haberl, M.D., Department of Neurology, Klinikum Grosshadern, Marchioninistrasse 15, D-81377 Munich, Germany.

Acta Neurochir (1993) [Suppl] 59: 69–73

Assessment of Cerebral Haemodynamics in Comatose Patients by Laser Doppler Flowmetry – Preliminary Observations

R. Steinmeier, I. Bondar, and **C. Bauhuf**

Department of Neurosurgery, University of Erlangen-Nürnberg, Federal Republic of Germany

Summary

Preliminary observations of laser Doppler flowmetry (LDF) signal changes in comatose patients with severe head injury or higher grade subarachnoid haemorrhage are presented. The data demonstrate that LDF measurements by no means lend themselves to a straightforward and unequivocal interpretation. Three main sources of LDF signal "bias" are tentatively distinguished: (1) properties inherent to the LDF measurement system and (2) spatial as well as (3) temporal heterogeneity of functional responses of the cerebral microvascular bed lead to unexpected, unpredictable and seemingly "paradox" patterns of the LDF signals. Despite the exploratory character of our data, we are convinced that they strongly suggest a more critical and cautious appraisal of the present possibilities of LDF in monitoring comatose patients than suggested by several other recent reports on this topic.

Keywords: Laser Doppler flowmetry; head injury; subarachnoid haemorrhage; critical appraisal.

Introduction

Laser Doppler flowmetry (LDF) is one of the most recently developed methods for continuously estimating local microvascular tissue perfusion and is increasingly used in a large variety of different organs and diseases. Applications in the field of neurosurgery range from animal investigations[7, 11, 13, 14, 19] to intraoperative and bedside monitoring of regional cerebral blood flow (CBF)[1, 12, 22, 26]. Several studies were performed using various cerebral preparations in different animal models to validate the method by correlation of the LDF measurements with simultaneous measurements by radionuclide labelled microspheres[11, 19], H_2 clearance[14, 25] and autoradiography[10]. In most cases these investigations showed a fairly good correlation between estimates of relative changes in CBF measured with LDF versus the data obtained by more established methods. Comparisons between various techniques must nevertheless be regarded with great caution, because all of the above mentioned methods are open to criticism due to their limited accuracy caused by poor spatial and/or temporal resolution and by the heterogeneity inherent to the cerebral circulation[25]. Despite this, the above mentioned investigations suggested the feasibility of continuous long term monitoring of CBF in patients using LDF[22].

The aim of the project in progress since 1991 at the Intensive Care Unit (ICU) of our department is the development and application of a multiparametric monitoring system to record and analyse relevant systemic and local physiological parameters for early detection and, whenever possible, therapeutic control of cerebral metabolic and haemodynamic deterioration in comatose patients.

In this paper, several preliminary observations of LDF signal changes occurring in comatose patients with severe head injury and higher grade subarachnoid haemorrhage (SAH) are presented. The examples were carefully selected with the purpose of demonstrating that LDF measurements by no means lend themselves to a straightforward and unequivocal interpretation, as implicitly stated by other reports[12, 22].

Material and Methods

Laser Doppler Flowmetry and Data Acquisition

Laser Doppler flowmetry was performed with two Laserflo[R] blood perfusion monitors (Type BPM 403 A; Vasamedics, St. Paul, MN) using standard implantable disc probes (Type PD 434; 18 mm). The laser light (wave length = 780 nm, power approximately 2 mW) generated by the laser diode (gallium-aluminium-arsenide semiconductor) is delivered by means of a flexible optical

50 μm graded index fiber to the tissue, where it is partially absorbed and scattered by moving red blood cells (RBCs) and static tissue structures. The tissue from which back-scattered light is sampled is assumed to have a volume of approximately 1 mm³. Scattering at nonmoving tissue components will have no influence on the light frequency, whereas the remaining fraction of the back-scattered light is Doppler shifted by moving RBCs. The fraction of the back-scattered light sampled at the probe tip containing both the output and the Doppler-shifted frequencies is guided by two optical 100 μm graded index fibers to a photo-detector. After the signal has been cleared by bandpass filtering, its frequency and the magnitude of the fluctuating fraction of the photocurrent are transformed by a signal processor to the output signals *velocity* (VEL) (range 0–8 [kHz]), *fractional volume of the RBCs* (VOL) (range 0–1.6 [V/V]), *microflow* ([FLUX]) (scaled product of VEL and VOL, range 0–400 arbitrary units) and the *direct current voltage offset* (DC) (range 0–2.5 [V]). The sampling time constant was set to 0.1 seconds. All four LDF output signals from both devices were monitored and recorded simultaneously with the *electrocardiogram* (ECG) [V], *systemic arterial blood pressure* (aBP) [mmHg], *dynamic exspiratory pCO₂* (pCO₂^{exsp}) [mmHg], *intracranial pressure* (ICP) [mmHg] and *mean blood flow velocity* (MV) [cm/s] of the middle cerebral artery, as measured by the transcranial Doppler ultrasound technique. The analogue output signals were digitalized by a 16 channel, 12 bit resolution, 25 kHz throughput analog to digital converter (Type DT 2814; Data Translation Inc., Marlboro, Massachusetts). We used a specially designed data acquisition program on a personal computer which allowed control of the sampling rate and the duration of the sampling interval. The time-series presented in this paper were sampled at a rate of 4 Hz for varying time intervals ranging from 30 seconds up to 15 minutes.

LDF Probe Placement

The two LDF probes were introduced directly through the skin incision and a frontal burr hole placed for routine external ventricular drainage implantation. To minimize bending and facilitate introduction of the probes, the edges of the burr holes were slanted. The probes were gently pushed forward and placed subdurally over the fronto-polar and fronto-lateral pial cortical surface respectively at a distance of some centimeters from the burr hole. After intra-operative control and correction of the probe position until artifact-free LDF signals could be detected, the skin was closed with interrupted sutures, the probes being brought out between two of these sutures. In order to prevent undesired displacements, the probes were secured with sutures to the skin. Following completion of the study the probes were removed in the ICU through the scalp incision.

Patient Population and Measurement Protocol

The studies were performed in 8 comatose patients with severe head injury or higher grade SAH. The above mentioned systemic and local physiological parameters were monitored three times a day during an interval of up to several hours, the entire study lasting maximally five days. A CO₂-test was performed once a day to assess CO₂-reactivity of the cerebrovascular bed.

Results

A preliminary and exploratory analysis of the simultaneous measurements of systemic and local

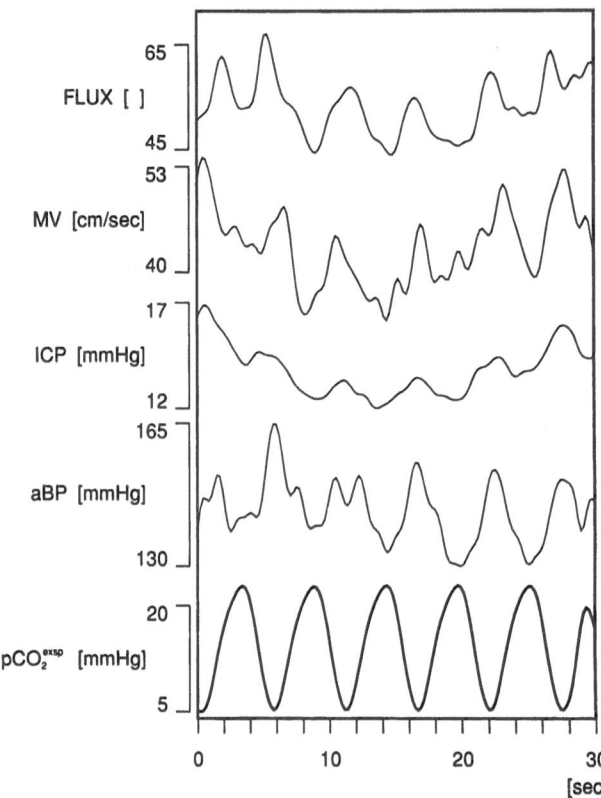

Fig. 1. Influence of the respiratory pattern as indicated by the pCO₂^{exsp} on cyclical changes of the LDF. *ICP, aBP* and mean blood flow velocity (*MV*) in the middle cerebral artery are recorded simultaneously

physiological parameters in comatose patients often revealed well-known and therefore expected, predictable and explainable interdependencies between these parameters. An example of such a relationship is given in Fig. 1, where the influence of the cyclic respiratory pattern (as indicated by pCO₂$^{exsp.}$) on several other parameters (aBP, ICP, MV and FLUX) can easily be discerned.

However, a closer look at the LDF signals showed some types of highly unexpected and seemingly "paradox", but perhaps explainable, patterns of the LDF signals, which we tentatively subdivided into three distinct categories:

(1) Fairly frequently we observed *"spontaneous", abrupt fluctuations of the LDF signals* without any simultaneous or shortly preceding change of other monitored parameters. These unpredictable "LDF signal drifts" were probably due to slight displacements of the LDF probe over the cerebral cortex, causing a change in the distance between the LDF probe and the brain surface and thus of the signal intensity. This interpretation is strengthened by the

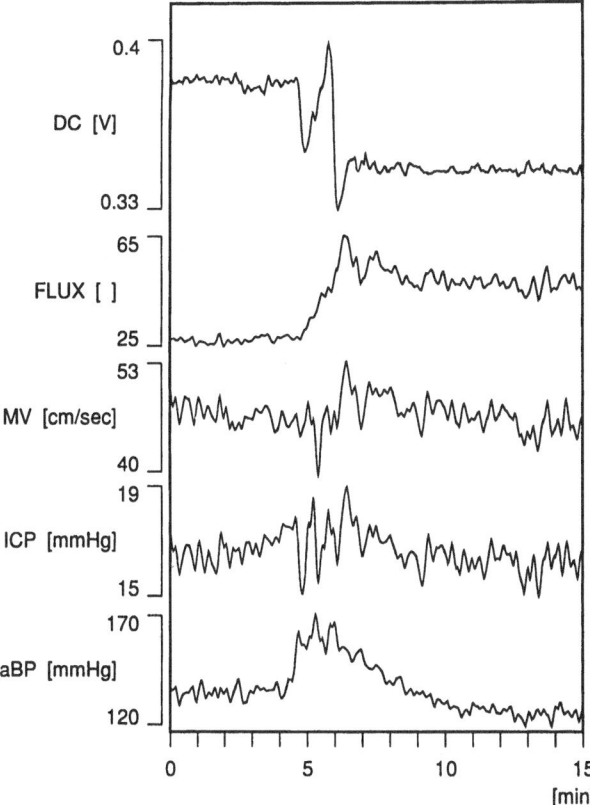

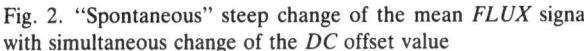

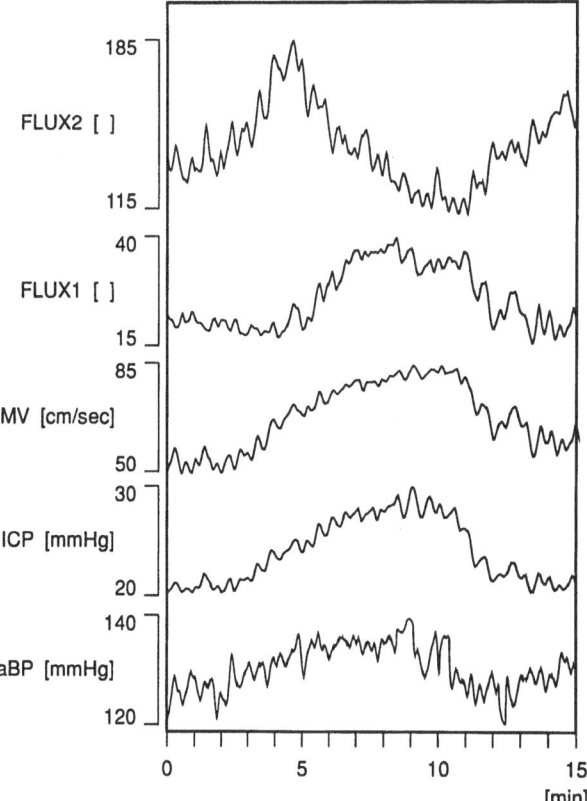

Fig. 2. "Spontaneous" steep change of the mean *FLUX* signal with simultaneous change of the *DC* offset value

Fig. 3. Spontaneous fluctuations in diametrically opposite directions of simultaneously recorded mean *FLUX* signals

marked change observed in the DC offset value (Fig. 2), which is directly proportional to the distance between the LDF probe and the brain surface[2]. Most intriguing is the fact that these fluctuations did not only occur during manipulations performed on the patient (e.g. endotracheal suctioning), but even under apparently "controlled" conditions, when the patient seemed to be at rest and no manipulations were carried out.

(2) When we looked at the simultaneous measurements of the same LDF parameter performed with two laser Doppler probes placed at different sites, "paradox" patterns with the LDF signals out of phase could often be discerned. The "spontaneous" change of the first signal in one direction was associated with a marked change of the corresponding second signal in the opposite direction (Fig. 3), suggesting *spatial heterogeneity of functional responses of the cerebral microvascular bed*.

(3) Another striking feature was the onset of signal changes during CO_2-tests in the expected direction but at different points in time (Fig. 4),

suggesting additional *temporal heterogeneity of functional responses of the cerebral microvascular bed*.

Spatial and temporal functional heterogeneity of the cerebral microvascular bed are thus at the root of the seemingly *"paradox" and "uncorrelated" patterns of simultaneously monitored LDF signals*.

Discussion

Continuous monitoring of global, regional and local cerebral blood flow in comatose patients with severe head injury, higher grade SAH and other conditions leading to diffuse or focal brain damage, raised intracranial pressure and impaired cerebral circulation would be of paramount importance in improving treatment strategies and consequently clinical outcome in these patients. Despite various technological advances in the last decades, unfortunately there still is no straightforward approach to a simple, reliable and continuous assessment of global and regional changes of the brain macro- and microcirculation. The methods available for

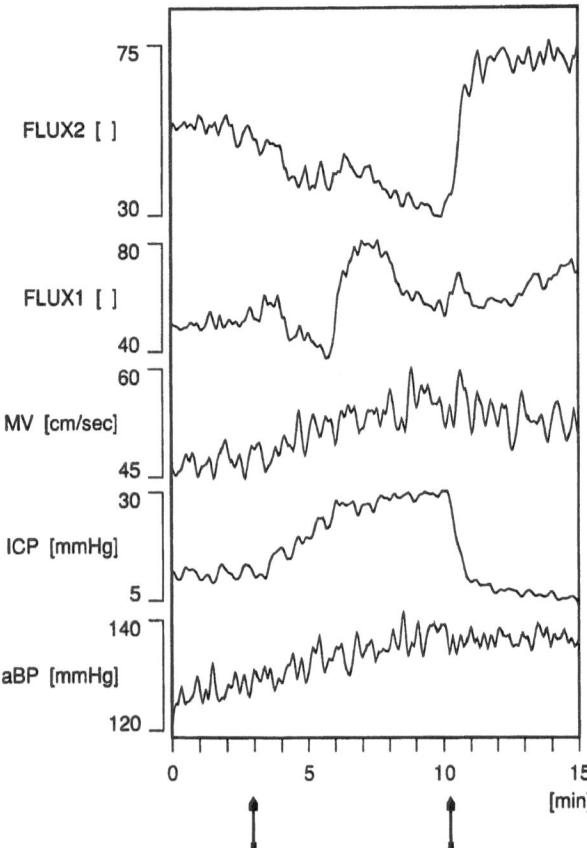

Fig. 4. Fluctuations of two simultaneously recorded mean *FLUX* signals being out of phase during a CO_2-test; begin and end of the CO_2-test are marked by arrows

monitoring of physiological parameters related to the actual state of the cerebral macro- and microcirculation in humans, such as xenon$_{133}$[3–5, 15–16, 21] and thermal clearance[8], thermal diffusion[9, 18, 27], transcranial Doppler sonography[6, 17, 21, 23, 24, 28] and imaging procedures such as single photon emission CT (SPECT) and xenon enhanced CT[20] are sometimes cumbersome and have either limited spatial or temporal resolution. For these reasons none of the above mentioned methods has entered or is on the threshold of entering into routine clinical application. LDF as one of the newer tools for continuous local blood flow measurement promised to provide useful information about local cerebral blood perfusion. It is therefore increasingly used in a wide variety of clinical and experimental settings in the field of neurosciences.

In this paper we present preliminary observations made during the multiparametric monitoring of systemic and local physiological paramaters in comatose patients on our ICU. The examples of LDF measurements presented were selected to demonstrate the highly unpredictable and "paradox" paths the signals often actually take, beside the fact that "well behaved" signal patterns can be isolated by appropriate data selection.

In our opinion, these "inconsistent" LDF signal patterns are probably caused by three main sources of "bias": (1) specific properties of the LDF measurement system as applied in the described clinical setting lead to slight displacements of the LDF probe and consequently to more or less marked and abrupt changes of the LDF signals. Despite several efforts to improve the stability of the probe position, we could not manage to avoid this major source of error for longer measurements. More important, the data strongly suggest the presence of (2) spatial and (3) temporal heterogeneity of functional responses of the microvascular bed. Postulating the existence of a microvascular module[29] (i.e. a local functional unit of the microvasculature), these observations strongly suggest a relative autonomy of the "modules" of the cerebral microvascular bed.

The above mentioned properties inherent in the measurement setup and the physiological system investigated severely compromise a straight-forward and unequivocal interpretation of the information provided by LDF measurements of local cerebral perfusion.

There is a distinct danger of severely discrediting potentially useful clinical applications of LDF by some enthusiasts of this new tool. It may well be that LDF will one day offer "a unique opportunity to monitor easily rCBF in pathological states and accurately reflect changes in regional flow"[22], but so far we do not think it does. Therefore the establishment of LDF as a valuable tool for understanding the pathophysiology and improve the management of comatose patients will require a great deal of further experimental and clinical investigations as well as theoretical reflection in order to better understand the complex systems involved both in the regulation of cerebral haemodynamics and in the interaction between the LDF device and brain tissue.

References

1. Arbit E, Di Resta GR, Bedford RF, Shah NK, Galicich JH (1989) Intraoperative measurement of cerebral and tumor blood flow with laser-doppler flowmetry. Neurosurgery 24: 166–170

2. Bondar I, Ronneberger D, Sigwanz U, Steinmeier R (1992) Distance between LDF-probe and brain surface of anesthetized rabbits affects LDF signals (abstract). Int J Microcirc Clin Exp 11 [Suppl 1]: 16

3. Bouma GJ, Muizelaar JP (1990) Relationship between cardiac output and cerebral blood flow in patients with intact and with impaired autoregulation. J Neurosurg 73: 368–374

4. Bouma GJ, Muizelaar JP, Bandoh K, Marmarou A (1992) Blood pressure and intracranial pressure-volume dynamics in severe head injury: relationship with cerebral blood flow. J Neurosurg 77: 15–19

5. Bouma GJ, Muizelaar JP, Choi SC, Newlon PG, Young HF (1991) Cerebral circulation and metabolism after severe traumatic brain injury: the elusive role of ischemia. J Neurosurg 75: 685–693

6. Chan K-H, Miller JD, Dearden NM, Andrews PJD, Midgley S (1992) The effect of changes in cerebral perfusion pressure upon middle cerebral artery blood flow velocity and jugular bulb venous oxygen saturation after severe brain injury. J Neurosurg 77: 55–61

7. Chen ST, Hsu CY, Hogan EL, Maricq H, Balentine JD (1986) A model of focal ischemic stroke in the rat: reproducible extensive cortical infarction. Stroke 17: 738–743

8. Dernbach PD, Little JR, Jones SC, Ebrahim ZY (1988) Altered cerebral autoregulation and CO_2 reactivity after aneurysmal subarachnoid hemorrhage. Neurosurgery 22: 822–826

9. Dickman CA, Carter LP, Baldwin HZ, Harrington T, Tallman D (1991) Continuous regional cerebral blood flow monitoring in acute craniocerebral trauma. Neurosurgery 28: 467–472

10. Dirnagl U, Kaplan B, Jacewitz M, Pulsinelli W (1989) Continuous measurement of cerebral cortical blood flow by laser-Doppler flowmetry in a rat stroke model. J Cereb Blood Flow Metab 9: 589–596

11. Eyre JA, Essex TJH, Flecknell PA, Bartholomew PH, Sinclair JI (1988) A comparison of measurements of cerebral blood flow in the rabbit using laser Doppler spectroscopy and radionuclide labelled microspheres. Clin Phys Physiol Meas 9: 65–74

12. Fasano VA, Urciuoli R, Bolognese P, Mostert M (1988) Intraoperative use of laser Doppler in the study of cerebral microvascular circulation. Acta Neurochir (Wien) 95: 40–48

13. Haberl RL, Heizer ML, Ellis EF (1989) Laser-Doppler assessment of brain microcirculation: effect of local alterations. Am J Physiol 256: H1255–H1260

14. Haberl RL, Heizer ML, Marmarou A, Ellis EF (1989) Laser-Doppler assessment of brain microcirculation: effect of systemic alterations. Am J Physiol 256: H1247–H1254

15. Heilbrun MP, Olesen J, Lassen NA (1972) Regional cerebral blood flow studies in subarachnoid hemorrhage. J Neurosurg 37: 36–44

16. Jaggi JL, Obrist WD, Gennarelli TA, Langfitt TW (1990) Relationship of early cerebral blood flow and metabolism to outcome in acute head injury. J Neurosurg 72: 176–182

17. Klingelhöfer JK, Sander D, Holzgraefe M, Bischoff C, Conrad B (1991) Cerebral vasospasm evaluated by transcranial Doppler ultrasonography at different intracranial pressures. J Neurosurg 75: 752–758

18. Kuwayama N, Takaku A, Harada J, Fukuda O, Endo S, Saito T (1991) Modified thermal diffusion flow probe for the continuous monitoring of cortical blood flow. Neurosurgery 29: 583–589

19. Lindsberg PT, O'Neill JT, Paakari IA, Hallenbeck JM, Feuerstein G (1989) Validation of laser-Doppler flowmetry in measurement of spinal cord blood flow. Am J Physiol 257: H674–H680

20. Marion DW, Darby J, Yonas H (1991) Acute regional cerebral blood flow changes caused by severe head injuries. J Neurosurg 74: 407–414

21. Martin NA, Doberstein C, Zane C, Caron MJ, Thomas K, Becker DP (1992) Posttraumatic cerebral arterial spasm: transcranial Doppler ultrasound, cerebral blood flow, and angiographic findings. J Neurosurg 77: 575–583

22. Meyerson BA, Gunasekera L, Linderoth B, Gazelius B (1991) Bedside monitoring of regional cortical blood flow in comatose patients using laser Doppler flowmetry. Neurosurgery 29: 750–755

23. Seiler RW, Nirkko AC (1990) Effect of nimodipine on cerebrovascular response to CO_2 in asymptomatic individuals and patients with subarachnoid hemorrhage: a transcranial Doppler ultrasound study. Neurosurgery 27: 247–251

24. Sekhar LN, Wechsler LR, Yonas H, Luyckx K, Obrist W (1988) Value of transcranial Doppler examination in the diagnosis of cerebral vasospasm after subarachnoid haemorrhage. Neurosurgery 22: 813–821

25. Skarphedinsson JO, Hårding H, Thorén P (1988) Repeated measurements of cerebral blood flow in rats. Comparisons between the hydrogen clearance method and laser Doppler flowmetry. Acta Physiol Scand 134: 133–142

26. Steinmeier R, Fahlbusch R, Powers AD, Dötterl A, Buchfelder M (1991) Pituitary microcirculation: physiological aspects and clinical implications. A laser-Doppler flow study during transsphenoidal adenomectomy. Neurosurgery 29: 47–54.

27. Tenjin H, Hirakawa K, Mizukawa N, Yano I, Ohta T, Uchibori M, Hino A (1988) Dysautoregulation in patients with ruptured aneurysms: cerebral blood flow measurements obtained during surgery by a temperature-controlled thermoelectrical method. Neurosurgery 23: 705–509

28. Weber M, Grolimund P, Seiler RW (1990) Evaluation of posttraumatic cerebral blood flow velocities by transcranial Doppler ultrasonography. Neurosurgery 27: 106–112

29. Zweifach BW (1986) Biomechanics of the microcirculation. In: Schmid-Schoenbein GW et al (eds) Frontiers in biomechanics. Springer, Berlin Heidelberg New York Tokyo, pp 283–298

Correspondence and Reprints: Dr. R. Steinmeier, Neurochirurgische Klinik der Universität Erlangen-Nürnberg, D-91054 Erlangen, Federal Republic of Germany.

Acta Neurochir (1993) [Suppl] 59: 74–80

Near Infrared Spectroscopy

Measurement of Adult Cerebral Haemodynamics Using Near Infrared Spectroscopy

C. E. Elwell, H. Owen-Reece[1], M. Cope, J. S. Wyatt[1], A. D. Edwards[1], D. T. Delpy, and E. O. R. Reynolds[1]

Departments of Medical Physics and Bioengineering and [1]Paediatrics, University College London, London, U.K.

Summary

Near infrared spectroscopy (NIRS) is a non invasive, portable, safe technique for monitoring cerebral oxygenation and haemodynamics. Since it does not involve the use of ionising radiation it may be used repeatedly to produce serial measurements of CBF and CBV in patients, and continuously to provide trend data about cerebral circulation changes. NIRS allows measurements to be made at the bedside with minimal disturbance to other monitoring and treatment procedures. Although regional information is not yet available, good time resolution allows rapid changes in cerebral haemodynamics to be observed.

Keywords: Near infrared spectroscopy; cerebral haemodynamics; cerebral oxygenation.

Introduction

Since it was first described in 1977[30], the technique of near infrared spectroscopy (NIRS) has gained widespread recognition as a non invasive clinical tool for the monitoring of changes in cerebral oxygenation[3, 4, 21, 22, 27, 42]. The majority of early NIRS studies were qualitative in nature, but recently technological and methodological developments have made it possible to quantify NIRS data. This paper will review the theory of NIRS, its instrumentation, and its application to the quantification of cerebral haemodynamics in adults.

Methods

Theory

Near infrared spectroscopy depends upon the relative transparency of biological tissue to light in the near infrared region of the spectrum. Light at visible wavelengths (450–700 nm) is strongly attenuated in tissue and as a result can only penetrate a maximum distance of approximately 1 cm. However, the absorption of light by the tissue chromophores haemoglobin and myoglobin is significantly lower at near infrared wavelengths (700–1 000 nm), and with sensitive detectors it is then possible to detect light which has traversed up to 8–9 cm of tissue, thus allowing measurements to be made in the adult as well as in the neonatal head[9, 18, 23, 40].

When light enters tissue it is both scattered and absorbed; the amount of each depends upon the wavelength of the light and the tissue type[8]. In any tissue the majority of the light attenuation due to scatter will be constant as long as the geometry does not alter. Similarly, attenuation due to light absorption by chromophores at a fixed concentration will remain constant. There are, however, three compounds which are present at variable concentrations in the brain and whose absorption characteristics vary with their oxygenation status. These are oxyhaemoglobin (HbO_2), deoxyhaemoglobin (Hb), and oxidised cytochrome aa₃ ($CytO_2$) – the terminal member of the respiratory chain. When near infrared light is shone through tissue containing these chromophores, the changes in their concentration can be quantified using a modified Beer Lambert Law, which describes optical attenuation in a highly scattering medium[11]. This can be expressed as:

$$\text{Attenuation (OD)} = \log\frac{I_o}{I} = \alpha c L B + G \qquad (1)$$

where OD represents optical densities, I_o the incident light intensity, I the detected light intensity, α the absorption coefficient of the chromophore ($mM^{-1}.cm^{-1}$), c the concentration of chromophore (mM), L the physical distance between the points where light enters and leaves the tissue (cm), B a "pathlength factor" which takes into account the scattering of light in the tissue, and G a factor related to the measurement geometry and type of tissue. If measurements are only made of the changes in attenuation, then L, B and G can be assumed to remain constant and changes in chromophore concentration can be derived from the expression:

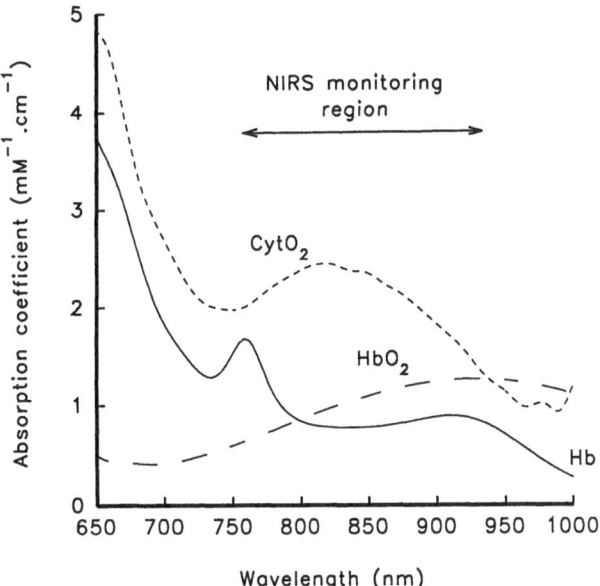

Fig. 1. Absorption spectra of oxyhaemoglobin (*HbO₂*), deoxy-haemoglobin (*Hb*), and oxidised cytochrome oxidase (*CytO₂*)

$$\delta c = \frac{\delta OD}{\alpha LB} \quad (2)$$

The absorption coefficients of HbO_2 and Hb have been measured in pure haemoglobin solutions, while those of cytochrome have been obtained from in vivo studies on fluorocarbon exchange-transfused rats[10, 41]. These absorption spectra are shown in Fig. 1. The pathlength factor, B, has been determined by measurement of the time of flight of ultra short light pulses through tissue, and for the adult head has a mean value of $5.93 \pm$ S.D. 0.42^{39}.

NIRS measurements cannot be made in transillumination across the adult head because of its diameter and the consequent degree of light attenuation across it. Partial transmission spectroscopy is therefore performed with the NIRS fibres at some acute angle to each other on the head. The validity of the modified Beer Lambert relationship with this geometry has been demonstrated both by experimental measurements and computer modelling of light interaction with tissue[37–39].

Instrumentation

The NIR spectrometer used in the clinical studies reported here (NIRO 500, Hamamatsu Photonics, Japan) is a commercial version of an instrument originally designed and built at the Medical Physics Department, University College London, details of which have already been published[10]. Briefly, the unit uses four pulsed laser diodes as light sources, their emission wavelengths being between 750 and 920 nm, and a photomultiplier tube for light detection. The NIR light is carried to and from the spectrometer through flexible fibre optic bundles, the patient end of which terminates in a small (<2 cm diameter) cylindrical optode. In studies of the adult brain, the optodes are placed high on one side of the forehead, away from the midline air and cerebral venous sinuses, and the temporalis muscle. The sinuses are avoided since light can be channelled by these away from the brain, whilst positions near the temporalis muscle may lead to significant contamination of the NIRS signal by surface tissues[28]. The exact positioning of the optodes is clearly dependent upon the level of the hairline of the subject but they are typically placed

at an acute angle to each other with an interoptode spacing of 4–7 cm. They are held in position using double sided adhesive rings and self adhesive tape. The head is then wrapped in a light proof cloth to reduce background light and improve the signal to noise ratio. Secure attachment of the optodes is important since small movements leading to changes in the interoptode spacing may be falsely interpreted as a change in chromophore concentration.

Using a previously derived algorithm[9] incorporating the absorption characteristics of the three chromophores, the measured changes in attenuation at each wavelength can be converted into equivalent changes in the concentrations of HbO_2, Hb, and $CytO_2$. All measurements are expressed as absolute concentration changes from an arbitrary zero at the start of the measurement period. With a knowledge of the pathlength factor B and the interoptode spacing, these changes can be quantified in $\mu mol.l^{-1}$. Alterations in the total cerebral haemoglobin concentration (Hb_{sum}) may be calculated from the sum of $\Delta[HbO_2]$ and $\Delta[Hb]$ and the difference between the signals ($\Delta[HbO_2] - \Delta[Hb]$) indicates the changes in cerebral haemoglobin oxygenation (Hb_{diff}).

In the NIRO 500 the power density of the light at the surface of the optode is more than one order of magnitude below international safety standards for skin (IEC 825). The power levels are also less than the international safety limits for eye exposure for both "intrabeam" and "extended source" viewing.

The current NIRO 500 system is capable of measuring the chromophore concentration changes at a maximum rate of 2Hz. This has allowed rapid changes in cerebral oxygenation to be detected.

Quantitation of Cerebral Blood Flow

The NIRS measurement of CBF uses a modification of the Fick principle which states that the rate of accumulation of a tracer in an organ is equivalent to the difference between its rate of arrival (flow × arterial concentration) and rate of departure (flow × venous concentration). If measurements of tracer accumulation are made within the assumed minimum transit time of the organ (in this case the adult brain), then the venous concentration of the tracer is zero so the rate of departure of the tracer can be ignored. CBF can then be determined from the ratio of the amount of tracer accumulated at some time t, to the quantity of the tracer introduced over the same time.

In NIRS the tracer used is the near infrared "dye" oxy-haemoglobin. When a sudden increase is induced in fractional arterial haemoglobin saturation (ΔSaO_2, which can be measured using a pulse oximeter on the ear), the resulting increase in cerebral oxyhaemoglobin concentration ($\Delta[HbO_2]$ – measured by NIRS) represents the accumulation of the tracer. The actual quantity of tracer introduced can be calculated from the product of the integral of the change in SaO_2 during time t, and the concentration of haemoglobin in whole blood ([tHb] – measured from a venous blood sample in $g.100 ml^{-1}$). CBF can then be calculated from:

$$CBF \ (ml.100\,g^{-1}.min^{-1}) = \frac{K \cdot \Delta[HbO_2]}{([tHb] \cdot 10^{-2}) \cdot \int_0^t \Delta SaO_2 \, dt} \quad (3)$$

where

$$K = \frac{MW_{Hb} \cdot 10^{-3}}{Dt \cdot 10} \quad (4)$$

MW_{Hb} is the molecular weight of haemoglobin and Dt is the cerebral tissue density in $g.ml^{-1}$.

If the total tissue haemoglobin concentration (Hb_{sum}) during this period is constant then it may be assumed that the changes in [Hb] and [HbO$_2$] are equal and opposite. The signal representing the difference between [Hb] and [HbO$_2$] is then twice the amplitude of the corresponding signal representing the change in [HbO$_2$] alone. Use of this parameter improves the signal to noise ratio of the NIRS measurement, and equation (3) can then be modified to:

$$\text{CBF (ml.100 g}^{-1}.\text{min}^{-1}) = \frac{K \cdot \Delta[\text{HbO}_2] - \Delta[\text{Hb}]}{2 \cdot [\text{tHb}] \cdot \int_0^t \Delta \text{SaO}_2 \, dt}$$

$$= \frac{K \cdot \Delta\text{Hb}_{\text{diff}}}{2 \cdot [\text{tHb}] \cdot \int_0^t \Delta \text{SaO}_2 \, dt} \quad (5)$$

This method of measuring flow was initially applied to cerebral studies in neonates[12, 14] and has subsequently been validated by comparison with ^{133}Xenon[2, 36]. This method has also been used to measure blood flow in the adult forearm, and validated by comparison with strain gauge plethysmography[13].

Details of the methodological problems associated with measuring the relatively high cerebral blood flows in adults using this technique have already been documented[16]. As described above, a small rapid increase in SaO$_2$ must be induced by altering the subject's inspired oxygen fraction (FiO$_2$). This change in SaO$_2$ must be accurately monitored by a pulse oximeter, and the resulting change in cerebral [HbO$_2$] by NIRS. The relatively short adult cerebral vascular transit time requires that the measurements be made over a period of a few seconds. It is therefore essential to use an oximeter capable of working in beat to beat mode – collecting data with minimal signal averaging so that the rapid step change in SaO$_2$ can be accurately measured. Similarly, the NIRS system must have a high sampling rate to obtain a sufficient number of data points over which to calculate Δ[HbO$_2$] before the transit time is reached.

Since the oximeter and the NIRS spectrometer are monitoring oxygenation changes at different sites, there will be a time delay between the signals which must be taken into account in the data analysis. This delay can be minimised by choosing the ear as the oximetry site.

Quantitation of Cerebral Blood Volume

Cerebral blood volume can similarly be quantified using NIRS by inducing a small but slow change in arterial haemoglobin saturation (ΔSaO$_2$). The details of the theory of this measurement are described in detail elsewhere[43]. Briefly, if CBV, CBF, and oxygen consumption remain constant, then the consequent change in cerebral [HbO$_2$] measured by NIRS, is equivalent to the product of the total cerebral haemoglobin concentration and the fractional change in arterial saturation. The total haemoglobin concentration derived from this relationship can then be converted into cerebral blood volume in ml.100 g^{-1} using data on cerebral brain tissue density and large to small vessel haematocrit ratio.

Results and Discussion

Clinical Monitoring

NIRS has now been employed to monitor tissue oxygenation in a number of clinical areas[7, 20, 25, 26] but to date has been most widely used in monitoring

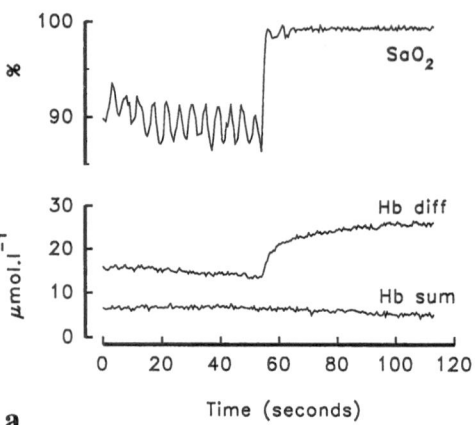

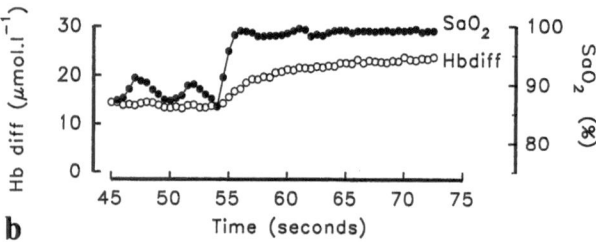

Fig. 2. (a) NIRS and SaO$_2$ data collected from a healthy subject during a CBF measurement and (b) the same data on an expanded scale showing more clearly the step change in SaO$_2$. Each point represents data averaged over 0.5 second

neonatal cerebral haemodynamics. Some early studies merely reported qualitative changes in oxygenation[1, 3, 4, 6, 19, 21], but recently quantitative measurements of CBF[2, 12, 36], CBV[32, 43] and its response to changes in carbon dioxide[35, 44] have been made. NIRS has also been used to quantify the changes in cerebral concentration of oxidised cytochrome oxidase in the newborn[15]. Changes in fetal cerebral oxygenation have been measured during labour and childbirth[33, 34]. Only recently has the quantification of adult cerebral haemodynamics been described[16].

Cerebral Blood Flow in Healthy Adults

Measurements of CBF have been made in ten healthy adults. Figure 2 shows representative data from a CBF measurement on one subject. The time period t, over which CBF was calculated, was 4 seconds. Figure 3 shows a plot of the tracer accumulated (Hb$_{\text{diff}}$) against the tracer introduced (ΔSaO$_2$). Since it is clear that there is not one constant rate of tracer accumulation, a polynomial has been fitted to these points. When this polynomial

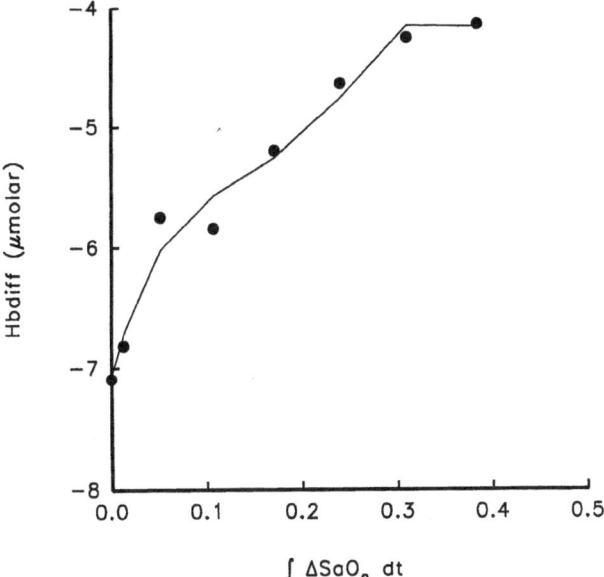

Fig. 3. The cumulative SaO_2 integral plotted against the equivalent change in Hb_{diff} from data shown in Fig. 2

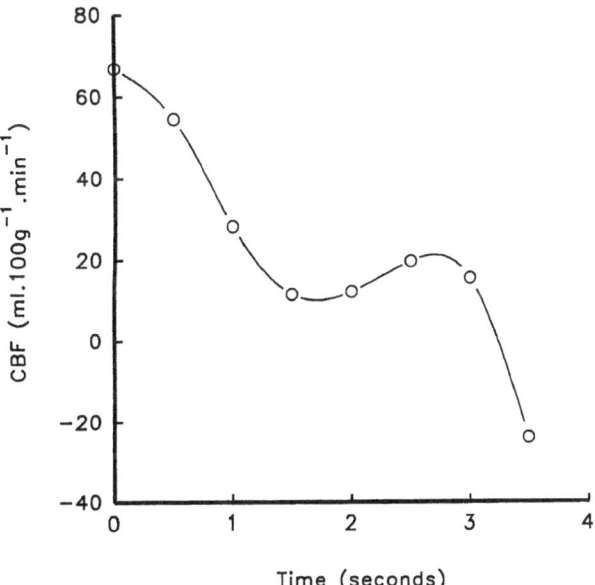

Fig. 4. The CBF versus time curve calculated from the data shown in Fig. 2. Note the negative flow value at the final point is an artefact of the mathematical method employed in the curve fitting procedure

is differentiated a CBF versus time curve is produced (Fig. 4), which clearly shows a high and low flow component. The pattern of the CBF versus time curve was reproducible over all ten subjects. The mean ($\pm$SD) high flow was 57 ± 19 ml.$100\,g^{-1}$.min^{-1} and the mean low flow component was 11 ± 4 ml.$100\,g^{-1}$.min^{-1}. These findings are similar to those

obtained with ^{133}Xenon clearance curves where it has been shown that at least two curves are needed to fit the data[24]. This has been interpreted as representing different flow rates in grey and white matter[45].

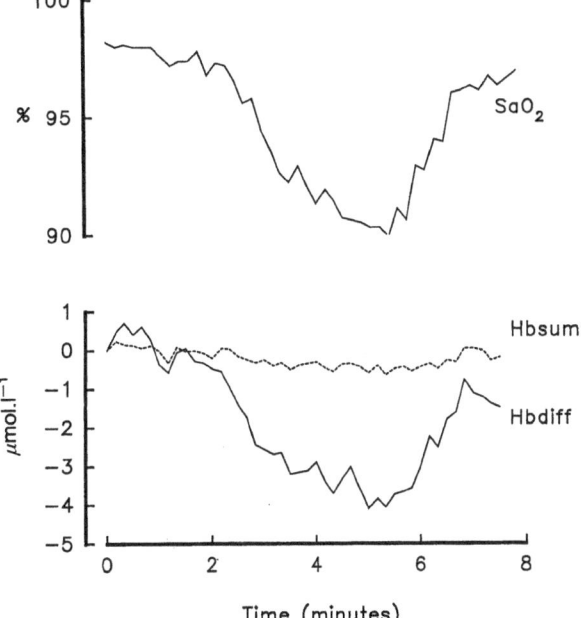

Fig. 5. NIRS and SaO_2 data collected from a healthy subject during a CBV measurement. A small transient reduction in inspired oxygen was induced over 5 minutes

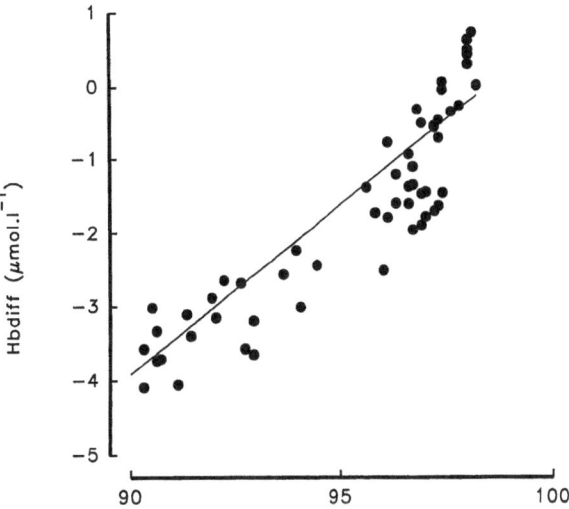

Fig. 6. Alteration in cerebral haemoglobin oxygenation (Hb_{diff}) plotted against SaO_2 for the data in Fig. 5. Cerebral blood volume is calculated from the slope of the regression line

Cerebral Blood Volume in Healthy Adults

Figure 5 shows the oximetry and NIRS data collected from one subject during a CBV measurement. FiO_2 was slowly reduced to produce a gradual fall in SaO_2 over approximately five minutes. If the change in cerebral oxygenation (Hb_{diff}) is plotted against the change in arterial saturation (ΔSaO_2) (Fig. 6), then the gradient of the regression line is proportional to CBV. CBV can be measured using an increase or decrease in SaO_2.

The mean CBV calculated from all ten subjects was $2.85 \pm 0.97 \, ml.100 \, g^{-1}$.

Contribution of Non-Cerebral Tissues

The contribution of surface tissues to the NIRS measurements in adults has not yet been quantified, although there is indirect evidence that surface contributions can be reduced by increasing the optode spacing[29]. In the neonate, the thinner surface tissue and skull, together with a smaller head diameter are thought to result in a small surface tissue contribution[31].

Intracranial NIRS measurements may be contaminated with either bone or scalp blood flow, or both. Given the relatively short time period over which the CBF measurements are made in adults and the relatively low values of bone blood flow[5], it is likely that the $[HbO_2]$ rise in the bone blood will not have commenced and thus bone blood flow will not be detected over this brief period. To investigate the contribution of scalp blood flow, some parallel CBF and CBV measurements were performed in subjects wearing a tourniquet inflated to arterial pressure around the base of the skull to occlude the scalp circulation. Similar values of CBF and CBV have been obtained pre and post occlusion (unpublished observations).

Experimental and theoretical data suggest that increasing the interoptode spacing increases the proportion of cerebral tissue interrogated. In measurement of the pathlength factor it is assumed that all tissues in the light path contribute to the $\Delta[HbO_2]$ and $\Delta[Hb]$ signals which are used to calculate CBF. If this assumption is invalid to whatever degree, there may be a certain percentage error in the quantitative CBF and CBV data.

Further Applications

The rapid sampling rate (2Hz) of the current NIRS system has allowed small respiratory linked oscillations in the cerebral circulation to be measured in normal volunteers breathing against an increased expiratory pressure. The changes detected suggest that homeostatic mechanisms do maintain CBF constant over the period of a single breath in healthy subjects[17]. It may therefore be possible to monitor the time course of the autoregulatory response using NIRS. Present clinical investigations include the quantification of CBF and CBV in patients undergoing intensive care. Studies are also planned to monitor the effects on the cerebral circulation of intraoperative procedures and anaesthesia.

Technological advances have made possible the development of a phase modulation system which will continuously measure the total pathlength which the light has travelled in the tissue. This will allow errors due to movement artefact to be accounted for and as such will be of particular use in NIRS measurements made on the fetus during labour and childbirth, and also in non sedated adults. Experimental and theoretical studies involving the modelling of light interaction with tissue are also underway to estimate the contribution of non cerebral tissue to NIRS signals from the adult head.

References

1. Benaron DA, Kurth CD, Steven J, Wagerle LC, Chance B, Delivoria-Papadopoulos M (1990) Cerebral oxygenation and oxygen consumption using phase shift spectrophotometry. Ann Inter Conference IEEE Engineering in Medicine and Biology Society 12: 2004–2006
2. Bucher HV, Edwards AD, Lipp AE, Duc G (1993) Comparison between ^{133}Xenon clearance and near infrared spectroscopy for estimation of cerebral blood flow. Paediatr Res 33: 56–60
3. Brazy JE, Lewis DV, Mitnick MH, Jöbsis FF (1985) Noninvasive monitoring of cerebral oxygenation in preterm infants: preliminary observations. Pediatrics 75: 217–225
4. Brazy JE, Lewis DV (1986) Changes in cerebral blood volume and cytochrome aa_3 during hypertensive peaks in preterm infants. Pediatrics 108: 983–987
5. Brookes M (1971) The blood supply to bone. Butterworth, London, p 236
6. Chance B, Smith DS, Nioka S, Miyoka H, Holton G, Maris M (1989) Photon migration in muscle and brain. In: Chance B (ed) Photon migration in tissues. Plenum Press, New York, pp 121–135
7. Cheatle TR, Potter LA, Cope M, Delpy DT, Coleridge-Smith PD, Scurr JH (1991) Near infrared spectroscopy in peripheral vascular disease. Br J Surg 78: 405–408
8. Cheong WF, Prahl SC, Welch AJ (1990) A review of the optical properties of biological tissues. IEEE J Quant Electron 26: 2166–2185
9. Cope M (1991) The development of a near infrared spectroscopy system and its application for non invasive monitoring of cerebral blood and tissue oxygenation in the newborn infant. PhD Thesis, University of London

10. Cope M, Delpy DT (1988) A system for long term measurement of cerebral blood and tissue oxygenation in newborn infants by near infrared transillumination. Med Biol Eng Comput 26: 289–294

11. Delpy DT, Cope M, van der Zee P, Arridge SR, Wray S, Wyatt JS (1988) Estimation of optical pathlength through tissue from direct time of flight measurement. Phys Med Biol 33: 1433–1442

12. Edwards AD, Wyatt JS, Richardson CE, Delpy DT, Cope M, Reynolds EOR (1988) Cotside measurement of cerebral blood flow in ill newborn infants by near infrared spectroscopy. Lancet ii: 770–771

13. Edwards AD, Reynolds EOR, Richardson CE, Wyatt JS (1988) Estimation of blood flow in man using near infrared spectroscopy (NIRS). J Physiol 410: 50P

14. Edwards AD, Wyatt JS, Richardson CE, Potter A, Cope M, Delpy DT, Reynolds EOR (1990) Effects of indomethacin on cerebral haemodynamics and oxygen delivery investigated by near infrared spectroscopy in very preterm infants. Lancet i: 1491–1495

15. Edwards AD, Brown GC, Cope M, Wyatt JS, McCormick DC, Roth SC, Delpy DT, Reynolds EOR (1992) Quantification of concentration changes in neonatal human cerebral oxidised cytochrome oxidase. J Appl Physiol 71: 1907–1913

16. Elwell CE, Cope M, Edwards AD, Wyatt JS, Reynolds EOR (1992) Measurement of cerebral blood flow in adult humans using near infrared spectroscopy - methodology and possible errors. Adv Exp Med Biol 317: 235–245

17. Elwell CE, Owen-Reece H, Cope M, Edwards AD, Wyatt JS, Reynolds EOR, Delpy DT (1992) Measurement of the changes in cerebral haemodynamics during inspiration and expiration in adults using near infrared spectroscopy. Adv Exp Med Biol (in press)

18. Essenpreis M, Spahn J, Waidelich W, Versmold HT (1990) Krypton filled flashlamp: a possible new light source for near infrared spectroscopy in vivo. Adv Exp Med Biol 277: 59–62

19. Faris F, Rolfe P, Thorniley M, Wickramasinghe Y, Houston R, Doyle M, O'Brien S (1992) Non invasive optical monitoring of cerebral blood oxygenation in the foetus and newborn: preliminary investigation. J Biomed Eng 14: 303–306

20. Ferrari M, Zanette E, Giannini I, Sideri G, Fieschi C, Carpi A (1986) Effects of carotid compression test on regional cerebral blood volume haemoglobin oxygen saturation and cytochrome-c-oxidase redox level in cerebrovascular patients. Adv Exp Med Biol 200: 213–222

21. Ferrari M, De Marchis, Giannini I, Nicola A, Agostino R, Nodari S, Bucci G (1986) Cerebral blood volume and haemoglobin oxygen saturation monitoring in neonatal brain by near infrared spectroscopy. Adv Exp Med Biol 200: 203–212

22. Fox EJ, Harme MH, Mitnick MH, Jobsis FF (1982) Non invasive monitoring of cerebral oxygen sufficiency during general anesthesia. Anesthesiology 57: A160

23. Giannini I, Ferrari M, Carpi A, Fasella P (1982) Rat brain monitoring by near infrared spectroscopy: an assessment of possible clinical significance. Physiol Chem Phys 14: 295–305

24. Ginsberg MD (1986) Cerebral circulation and its regulation and pharmacology. In: Asbury AK, McKhann GM, McDonald WI (eds) Diseases of the nervous system. WB Saunders, Philadelphia, pp 1074–1085

25. Greely WJ, Bracey VA, Ungerleider RM, Greibel JA, Kern FH, Boyd JL, Reves JG, Piantadosi CA (1991) Recovery of cerebral metabolism and mitochondrial oxidation state is delayed after hypothermic circulatory arrest. Circulation 84 [Suppl 5], III400–406

26. Hampson NB, Piantadosi CA (1988) Near infrared monitoring of human skeletal muscle oxygenation during forearm ischaemia. J Appl Physiol 64: 2449–2457

27. Hampson NB, Camporesi EM, Stolp BW, Moon RE, Shook JE, Greibel JA, Piantadosi CA (1990) Cerebral oxygen availability by NIR spectroscopy during transient hypoxia in humans. J Appl Physiol 69: 907–913

28. Harris DNF, Cowans FM, Wertheim DA (1992) NIRS in the temporal region – strong influence of external carotid artery. Adv Exp Med Biol (in press)

29. Harris DNF, Cowans FM, Wertheim DA, Hamid S (1992) NIRS in adults – effects of increasing optode separation. Adv Exp Med Biol (in press)

30. Jobsis FF (1977) Noninvasive infrared monitoring of cerebral and myocardial oxygen sufficiency and circulatory parameters. Science 198: 1264–1267

31. Kurth CD, Steven JM, Wagerle LC, Chance B, Delivoria-Papadopoulous M (1989) Bedside measurement of cerebral oxygenation by reflectance spectroscopy. [Abstract] Pediatr Res 25: 357

32. Livera LN, Spencer SA, Thorniley MS, Wickramasinghe Y, Rolfe P (1991) Effects of hypoxaemia and bradycardia on neonatal cerebral haemodynamics. Arch Dis Child 66: 376–380

33. Peebles DM, Edwards AD, Wyatt JS, Bishop AP, Cope M, Delpy DT, Reynolds EOR (1991) Effect of oxytocin on fetal brain oxygenation during labour. Lancet 338: 254–255

34. Peebles DM, Edwards AD, Wyatt JS, Bishop AP, Cope M, Delpy DT, Reynolds EOR (1992) Changes in human fetal cerebral haemoglobin concentration and oxygenation during labour measured by near infrared spectroscopy. Am J Obstet Gynecol 166: 1369–1373

35. Pryds O, Greisen G, Skov LL, Fris-Hansen B (1990) Carbon dioxide related changes in cerebral blood volume and cerebral blood flow in mechanically ventilated preterm neonates. Comparison of near infrared spectrometry and ^{133}Xenon clearance. Pediatr Res 27: 445–449

36. Skov L, Pryds O, Greisen G (1991) Estimating cerebral blood flow in newborn infants: comparison of near infrared spectroscopy and ^{133}Xenon clearance. Paediatr Res 30: 570–573

37. Tamura M, Nomura Y, Hazeki O (1987) Laser tissue spectroscopy – near infrared CT. Rev Laser Eng 15: 74–82

38. van der Zee P, Arridge SR, Cope M, Delpy DT (1990) The effect of optode positioning on optical pathlength in near infrared spectroscopy of brain. Adv Exp Med Biol 277: 79–84

39. van der Zee P, Cope M, Arridge SR, Essenpreis M, Potter LA, Edwards AD, Wyatt JS, McCormick DC, Roth SC, Reynolds EOR, Delpy DT (1992) Experimentally measured optical pathlengths for the adult head, calf and forearm and the head of the newborn infant as a function of interoptode spacing. Adv Exp Med Biol 316: 143–153

40. Wickramasinghe Y, Thorniley M, Rolfe P, Houston R, Livera N, Faris F (1990) Development of algorithms for non invasive neonatal cerebral monitoring using near infrared spectroscopy. Proc of 12th Ann Inter Conf IEEE Eng Biol Soc 12: 1554–1545

41. Wray S, Cope M, Delpy DT, Wyatt JS, Reynolds EOR (1988) Characterisation of the near infrared absorption spectra of cytochrome aa$_3$ and haemoglobin for the non invasive monitoring of cerebral oxygenation. Biochim Biophys Acta 933: 184–192

42. Wyatt JS, Cope M, Delpy DT, Wray S, Reynolds EOR (1986) Quantification of cerebral oxygenation and haemodynamics in sick newborn infants by near infrared spectrophotometry. Lancet 2: 1063–1066

43. Wyatt JS, Cope M, Delpy DT, Richardson CE, Edwards AD, Wray SC, Reynolds EOR (1990) Quantitation of cerebral blood volume in newborn infants by near infrared spectroscopy. J Appl Physiol 68: 1086–1091

44. Wyatt JS, Edwards AD, Cope M, Delpy DT, McCormick DC, Potter A, Reynolds EOR (1991) Response of cerebral blood volume to changes in arterial carbon dioxide tension in preterm and term newborn infants. Paediatr Res 29:553–557

45. Yonas H, Darby JM, Marks EC, Durham SR, Maxwell C (1991) CBF measured by Xe-CT: Approach to analysis and normal values. J Cereb Blood Flow Metab 11: 716–725

Correspondence and Reprints: C. E. Elwell, M. Phil., Department of Medical Physics and Bioengineering, University College London, 1st Floor Shropshire House, 11–20 Capper Street, London, WC1 6JA, U.K.

Acta Neurochir (1993) [Suppl] 59: 81–85

Transcranial Doppler Sonography

Transcranial Doppler-Sonography in Severe Head Injury

K.-H. Chan, N. M. Dearden, and **J. D. Miller**

Department of Clinical Neurosciences, University of Edinburgh, Western General Hospital, Edinburgh, Scotland, U.K.

Summary

Ischaemic brain damage is present in over 90% of patients suffering from fatal head injury. Early detection and treatment of ischaemia may improve outcome after head trauma. Monitoring of blood flow velocity of the middle cerebral artery by noninvasive transcranial doppler ultrasound provides an alternate means of identifying cerebral ischaemia.

Keywords: Severe head injury; ischaemic brain damage; blood flow velocity; transcranial doppler ultrasound; TCD.

Introduction

After brain injury, cerebral metabolism falls while cerebral blood flow (CBF) varies with depression of the level of consciousness. Secondary factors, including hypoxia, hypotension, and raised intracranial pressure (ICP), may further compromise CBF and oxygen delivery, resulting in cerebral ischaemia. Management of severe head injury should aim to prevent brain ischaemia by detecting and correcting decreased CBF and oxygen delivery to the brain. Intermittent methods of CBF measurements have been used in the past to define the pathophysiology of cerebral ischaemia[4, 5, 18]. However, these measurements cannot be performed in patients with rapidly changing haemodynamic status, like unstable ICP. Transcranial doppler (TCD) ultrasonography provides a non-invasive means of repeated or continuous recording of blood flow velocity in major basal intracranial vessels. Blood flow velocity is proportional to the ratio of regional CBF to the calibre of the vessel being examined. In theory, flow velocity is proportional to CBF if the size of the insonated vessel remains unchanged. Clinical and experimental evidences have shown that the relationship between blood flow velocity and CBF in non-linear[3, 13, 20]. The diameter of the major basal intracranial arteries may vary[7, 14]. Interpretation of flow velocity values in isolation as an index of CBF therefore requires caution. Recently, TCD has been used in the management of severe head injury as a means of detecting cerebral ischaemia.

Methods

TCD Monitoring of Patients with Severe Head Injury

TCD was evaluated as a monitoring tool in a group of 50 severely brain-injured patients[6, 9, 10]. The first study examined the relationship between changes in intracranial haemodynamic factors and alterations of TCD parameters in a subgroup of 41 patients with continuous ICP monitoring[10]. In the second study, the significance of increased TCD flow velocity to neurological status and outcome after head injury was defined in the entire group of 50 cases[6].

The clinical management has been described in detail previously[10]. All patients were treated by a standard protocol that included artificial ventilation under muscle paralysis and sedation. Blood pressure (BP), arterial oxygen saturation (SaO2), end-tidal carbon dioxide concentration, and body temperature were recorded continuously. 41 patients had ICP monitoring. 22 of these patients also had unilateral continuous monitoring of jugular bulb venous oxygen saturation (SJO2) using the Oximetrics 3 system (Abbott Laboratories, Chicago, Illinois, U.S.A.)[2, 11]. The arterial-jugular venous oxygen content difference (AVDO2) was calculated by multiplying the difference between SaO2 and SJO2 by the daily haemoglobin concentration and 1.39, divided by 100. Global cerebral hyperaemia was defined as AVDO2 of less than 4 ml/dl. Data obtained after ICP therapy were excluded from the present analysis.

Transcranial Doppler (Medasonics, California, U.S.A.) insonation of at least both middle cerebral arteries (MCA) was performed within 24 hours of admission according to the method described by Aaslid[1]. Subsequent daily measurements used the same window and depth previously employed. Continuous recordings of MCA velocity were obtained during periods of changing CPP by mounting the ultrasound probe in a head band. Velocity measurements used in this study were taken at the points of maximum and subsequent minimum CPP before treatment was instigated. TCD parameters measured included systolic (S), diastolic (D), and timed-mean (M) velocity, from which the pulsatility index (PI) was derived (S-D/M). MCA velocity was considered to be abnormally high if it exceeded 100 cm/second.

Results

Correlations between Cerebral Haemodynamic Factors and TCD Changes

In 41 patients with ICP monitoring, TCD changes were compared with alterations in ICP, BP, and CPP. As CPP decreased, either as a result of fall in BP or increase in ICP, there was a reduction in flow velocity with diastolic velocity falling more than systolic velocity resulting in an increase of PI.

A linear correlation was observed between ICP (r = −0.315, P < 0.001), BP (r = 0.271, p < 0.001), CPP (r = 0.477, p < 0.001), and timed-mean velocity. Similar correlations were noted between ICP (r = 0.439, p < 0.001), BP (r = −0.445, p < 0.001), CPP (r = −0.767, p < 0.0001), and PI. Thus, better correlation existed between CPP and PI. Sequential linear regression analysis of the PI/CPP curve showed a CPP breakpoint at 70 mmHg (Fig. 1).

In 22 patients with concomitant ICP and SJO2 monitoring, linear correlations were noted between SJO2 and ICP (r = −0.378, p < 0.001) and CPP (r = 0.685, p < 0.0001). As a result, SJO2 also correlated closely with CPP. Further analysis of the SJO2/CPP curve also revealed a CPP breakpoint value, at

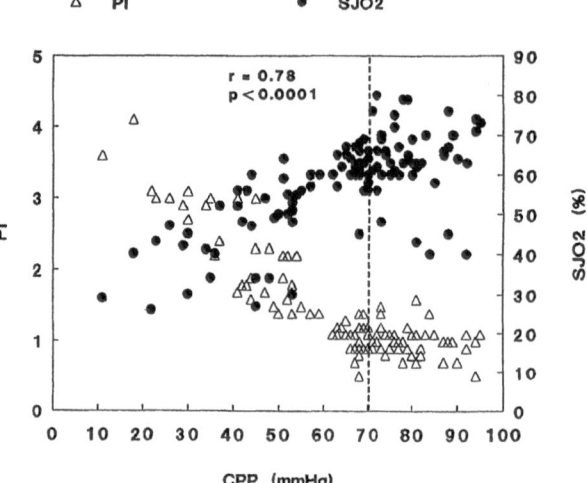

Fig. 2. Composite plot of SJO2 and PI versus CPP showing the same CPP breakpoint of 70 mmHg

71 mmHg. Below a CPP value of 71 mmHg, SJO2 progressively decreased as CPP fell (r = 0.78, p < 0.0001). This CPP breakpoint for SJO2 coincided with that of PI (Fig. 2).

Discussion

Factors Affecting TCD Measurements in Intensive Care

One of the major concerns in TCD recording is the variation in the angle between the ultrasound beam and the vessel being insonated during different measurements. Use of the dimensionless pulsatility index (PI) has the advantage of eliminating this potential error in measurement[10]. Continuous TCD monitoring also allows more reproducible recording than a hand-held probe because of the constancy of the ultrasound probe in relation to the vessel being examined[15]. Other factors which may affect serial TCD recordings include blood pressure, ICP, cerebral perfusion pressure (CPP), carbon dioxide level, arterial oxygen saturations, and haemoglobin concentration[10, 16].

Significance of PI and SJO2 Measurements

By Fick's equation, AVDO2 represents the ratio of cerebral metabolism to CBF. Since AVDO2 is derived from SJO2, SJO2 is a measure of the ratio of CBF to metabolism provided other factors as the position of the oxygen dissociation curve and haemoglobin concentration remain unchanged. If

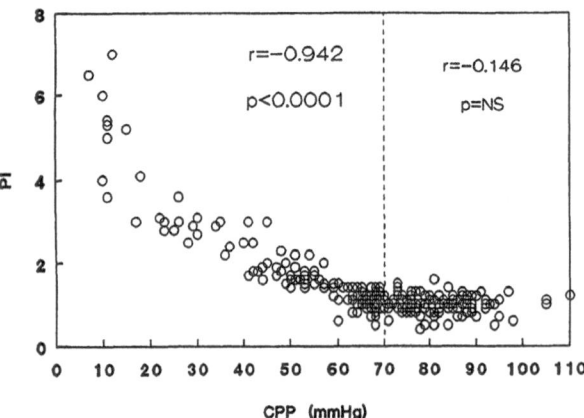

Fig. 1. Plots of PI versus CPP showing the CPP breakpoint of 70 mmHg

cerebral metabolism remains constant, a reduction in SJO2 as CPP decreases, indicates an increase in oxygen extraction by the brain to maintain its metabolic needs when CBF falls. A constant SJO2 suggests an unchanged CBF[10, 19].

The physiological goal of cerebral autoregulation is to maintain a level of CBF adequate to meet its metabolic demands[10, 19]. The lower limit of autoregulation is the threshold CPP value below which CBF adequate to meet tissue oxygen demands cannot be maintained. Our results suggest that continuous SJO2 monitoring allows identification of a threshold CPP level of 70 mmHg, above which CPP should be maintained during head injury management. Below the CPP breakpoint of 70 mmHg, the increase of PI with decreasing CPP represents progressive exhaustion of the mechanism of autoregulation and decrease of SJO2 represents increased oxygen extraction. Serial continuous TCD monitoring, like SJO2 measurements, may be used to define the lower limit of autoregulation. The CPP threshold of 70 mmHg identified in this study suggests that a higher CPP than previously recognised is required for the optimal management of patients with severe brain injury[17], and in individual cases even higher values may be required.

Correlation of Increased TCD Flow Velocity and Development of Ischaemic Neurological Complications

In the group of 50 patients described above, computerised tomography (CT) was performed on admission and whenever clinically indicated. Follow-up CT was obtained at three months in all survivors. Cerebral infarction was diagnosed from CT and post-mortem data. Infarction was classified into contusion-related (at or around the site of previous contusion or haematoma) and noncontusion related. The latter was identified as a hypodense area on CT with distinct margins restricted to an arterial territory, and persisted after the acute phase of injury.

Velocity exceeding 100 cm/second was observed in 17 of the 50 patients studied. All 17 patients had ICP recordings. Raised TCD velocity was noted only when CPP was above 60 mmHg.

Of the 17 cases with increased velocity, global cerebral hyperaemia based on AVDO2 values was associated with elevated velocity on six occasions.

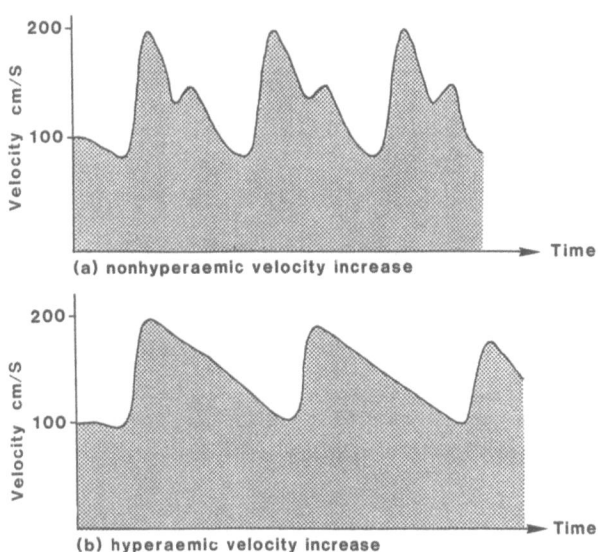

Fig. 3. (a) Diagram showing the notch in nonhyperaemic patient with increased velocity. (b) Diagram showing the absence of the notch in hyperaemic patients with velocity increase

The remaining patients were nonhyperaemic. In all six hyperaemic cases, increased velocity was present bilaterally in all vessels being examined. All non-hyperaemic cases except two had unilateral increase in MCA flow velocity. A characteristic waveform notch was noted at diastole in all nonhyperaemic patients with increased TCD velocity (Fig. 3). This was absent in the six hyperaemic cases[8].

Noncontusion related infarcts developed in four of the 17 patients with increased velocity. None of the 33 cases without raised velocity developed similar infarct. All four cases with noncontusion related infarcts had unilateral increased velocity in the territory of the relevant cerebral artery and were nonhyperaemic. None of the hyperaemic patients with increased velocity developed noncontusion related infarcts.

Therapy to elevate CPP to above 60 mmHg was required in 7 of 17 patients with increased velocity. All four patients who had elevated velocity and subsequently developed noncontusion related infarcts had received treatment to restore a reduced CPP. This finding suggested that patients with increased velocity who developed infarction might have had their lower limit of cerebral autoregulation set at a higher CPP level than those without infarcts. Applying the results of the first study, PI/SJO2/CPP plots can be used to define this lower autoregulatory limit in different patients. Figure 4 shows the PI/

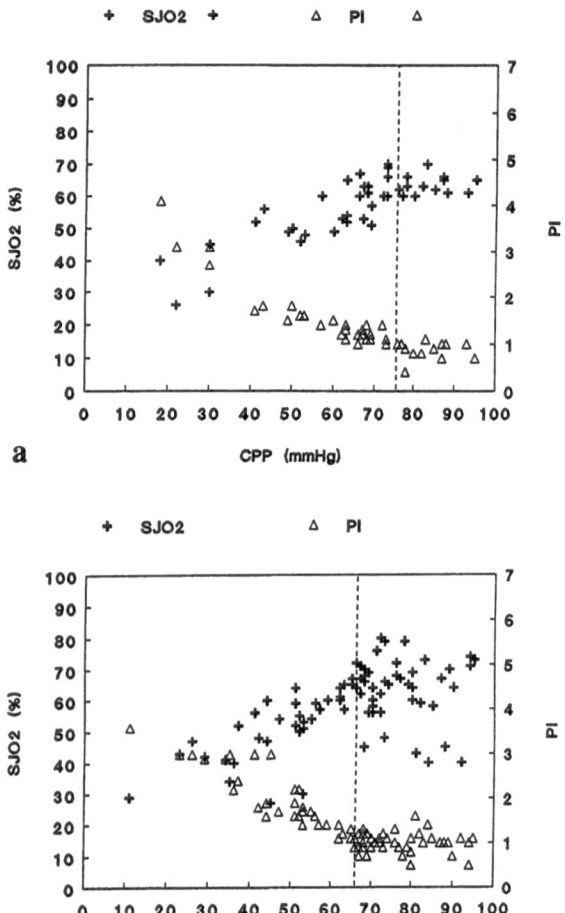

Fig. 4. PI/SJO2/CPP plots in patient with noncontusion related infarcts (a) showed a CPP breakpoint at 76 mmHg. A similar plot in patients without infarct (b) showed a CPP breakpoint at 67 mmHg

SJO2/CPP plots of these two groups of patients. The CPP breakpoints for patients with and without infarcts were respectively 76 and 67 mmHg.

These results demonstrate that documentation of increased TCD flow velocity requires an adequate CPP. Elevation of TCD flow velocity, especially if this is unilateral and accompanied by a reduction of CPP, significantly increases the risk of noncontusion related infarction.

Conclusions

Serial and continuous TCD recording is useful in defining the optimal CPP for management of individual severely head-injured patients. This critical CPP threshold may vary in different patients and at different time periods in the same patient. TCD may

be used to identify patients at risk of developing ischaemic infarction.

References

1. Aaslid R, Markwalder TM, Nornes H (1982) Noninvasive transcranial Doppler ultrasound recording of flow velocity in basal cerebral arteries. J Neurosurg 57: 769–774
2. Andrews PJD, Dearden NM (1990) Validation of the oximetrix III for continuous monitoring of jugular bulb oxygen saturation: comparison with IL282 in vitro co-oximeter. Br J Anaesth 64: 393–394
3. Bishop CCR, Powell S, Rutt D, Browse NL (1986) Transcranial doppler measurement of middle cerebral artery blood flow velocity: a validation study. Stroke 17: 913–915
4. Bruce DA, Langfitt TW, Miller JD, Schutz H, Vapalahti MP, Stanek A, Goldberg HI (1973) Regional cerebral blood flow, intracranial pressure, and brain metabolism in comatose patients. J Neurosurg 38: 131–144
5. Bouma GJ, Muizelaar JP (1990) Relationship between cardiac output and cerebral blood flow in patients with intact and with impaired autoregulation. J Neurosurg 73: 368–374
6. Chan KH, Dearden NM, Miller JD (1992) The significance of posttraumatic increase in cerebral blood flow velocity: A transcranial Doppler ultrasound study. Neurosurgery 30: 697–700
7. Chan KH, Dearden NM, Miller JD, Andrews PJD, Midgley S (1993) Multimodality monitoring as a guide to treatment of intracranial hypertension after severe brain injury. Neurosurgery (in press)
8. Chan KH, Dearden NM, Miller JD, Midgley S, Piper IR (1992) Transcranial doppler waveform differences in hyperaemic and nonhyperaemic patients after severe head injury. Surg Neurol 38: 433–436
9. Chan KH, Miller JD, Dearden NM (1992) Intracranial blood flow velocity after head injury: relationship to severity of injury, time, neurological status and outcome. J Neurol Neurosurg Psychiatry 55: 787–791
10. Chan KH, Miller JD, Dearden NM, Andrews PJD, Midgley S (1992) The effect of changes in cerebral perfusion pressure on middle cerebral artery blood flow velocity and jugular bulb venous oxygen saturation. J Neurosurg 77: 55–61
11. Dearden NM (1991) Jugular bulb venous oxygen saturation in the management of severe head injury. Curr Opin Anaesthiol 4: 279–286
12. Dickman CA, Carter LP, Baidwin HZ, Harrington T, Tallman D (1991) Continuous regional cerebral blood flow monitoring in acute craniocerebral trauma. Neurosurgery 28: 467–472
13. Halsey JH, McDowell HA, Gelmon S, Morawetz RB (1989) Blood velocity in the middle cerebral artery and regional cerebral blood flow during carotid endarterectomy. Stroke 20: 53–58
14. Heistad DD, Marcus ML, Abboud FM (1978) Role of large arteries in regulation of cerebral blood flow in dogs. J Clin Invest 62: 761–768
15. Lindegaard KF, Lundar T, Wiberg J, Sjoberg D, Aaslid R, Nornes H (1987) Variations in middle cerebral artery blood flow investigated with noninvasive transcranial blood velocity measurements. Stroke 18: 1025–1030
16. Markwalder TM, Grolimund P, Seiler RW, Roth F, Aaslid R (1984) Dependency of blood flow velocity in the middle cerebral artery on end-tidal carbon dioxide partial pressure – a transcranial ultrasound Doppler study. J Cereb Blood Flow Metab 4: 368–372

17. Miller JD (1985) Head injury and brain ischaemia. Implications for therapy. Br J Anaesth 57: 120–130
18. Obrist WD, Langfitt TW, Jaggi JL, Cruz J, Gennarelli T (1984) Cerebral blood flow and metabolism in comatose patients with acute head injury. Relationship to intracranial hypertension. J Neurosurg 61: 241–253
19. Paulson OB, Strandgaard S, Edvinsson L (1990) Cerebral autoregulation. Cerebrovasc Brain Metab Rev 2: 161–192
20. Rowan JO, Harper AM, Miller JD, Tedeschi GM, Jennett WB (1970) Relationship between volume flow and velocity in the cerebral circulation. J Neurol Neurosurg Psychiatry 33: 733–738

Correspondence and Reprints: K.-H. Chan, M.D., Division of Surgical Neurology, Department of Surgery, The University of Hong Kong, Queen Mary Hospital, Hong Kong.

Acta Neurochir (1993) [Suppl] 59: 86–90

Jugular-Venous Oximetry

Cerebral Oxygenation
Monitoring and Management

J. Cruz

Division of Neurosurgery and Head Injury Center, University of Pennsylvania, Philadelphia, Pennsylvania, U.S.A.

Summary

A comprehensive overview is presented on cerebral hemo-dynamics and metabolism in intensive care, with emphasis on cerebral oxygenation monitoring and management. From blood pressure to the most recently introduced concept of cerebral hemometabolism, cerebral hemodynamic reserve, concepts, variables, and findings are addressed. Integrated cerebral hemo-dynamics and metabolism, that is, cerebral hemometabolism, is discussed beyond cerebral perfusion pressure and blood flow. Monitoring and management of cerebral oxygenation and related variables are emphasized, in terms of current and potential future applications.

Keywords: Cerebral oxygenation; arteriojugular monitoring; intensive care.

Introduction

The human brain is highly dependent on adequate aerobic metabolism. This is clearly expressed by the fact that the brain, weighing only approximately 2% of the adult human mass, requires approximately 20% of the total body blood flow under normal physiologic conditions.

Cerebral hemometabolism was first investigated in 1942, by Gibbs and co-workers, in a large series of healthy human volunteers, most of whom medical students[15]. In this study, arteriojugular differences of oxygen and glucose (global cerebral extraction) and lactate (global cerebral production) were measured, under normocapnic conditions. Interestingly, at that time, the authors did not clearly state that cerebral arteriovenous differences of oxygen and glucose represented cerebral extraction of those elements, and that cerebral arteriovenous difference of lactate represented cerebral production of it. Nevertheless, they expressed that such measurements were somewhat representative of cerebral metabolism.

A few years later, Kety and Schmidt reported another pioneering contribution, for quantification of global cerebral blood flow (CBF)[17]. Because of this landmark contribution, it became possible to quantify brain oxygen consumption, or cerebral metabolic rate of oxygen consumption (CMRO$_2$). The basic standards of Kety and Schmidt's contribution are still routinely adopted; that is, blood flow is more frequently measured as height-over-area times the partition coeficient, and cerebral metabolic rate more frequently measured as the product of blood flow and arteriovenous difference (of any element).

Since intracranial pressure monitoring was introduced in humans[16], monitoring of cerebral perfusion pressure became a routine in many centers. Cerebral perfusion pressure, the difference between mean arterial pressure and intracranial pressure, has been advocated as a reliable measurement of cerebral hemodynamic adequacy[21, 24]. However, normal or even high cerebral perfusion pressure alone does not mean adequate cerebral blood flow relative to cerebral oxygen metabolic demand (CMRO$_2$). This is routinely found in ischemic stroke or vasospasm in subarachnoid hemorrhage, where in the presence of normal intracranial pressure, and normal or frequently high blood pressure (high cerebral perfusion pressure), profound cerebral ischemia is easily documented on clinical and subsidiary diagnostic tests.

The major limitation of cerebral perfusion pressure in informing about cerebral tissue perfusion adequacy relative to metabolism is because cerebral perfusion pressure is a gross estimate of cerebral hemodynamics, which does not allow an estimation of cerebral vascular resistance. Cerebral perfusion pressure is simply the difference of two pressures, which has absolutely no information on cerebral metabolism.

Cerebral blood flow is directly proportional to cerebral perfusion pressure, but inversely proportional to cerebral vascular resistance. Thus, cerebral blood flow is a more accurate measurement of cerebral hemodynamics than cerebral perfusion pressure. Yet, cerebral blood flow alone provides absolutely no information about the adequacy or inadequacy of itself to satisfy cerebral metabolic demand; that is, cerebral blood flow is also simply a hemodynamic (not a hemometabolic) variable, irrespectfully of the technique.

Therefore, the truly cerebral hemometabolic variables are the arteriovenous differences, as they accurately express the exchange between the capillary and the tissue. This implies that, for any values of cerebral blood flow and cerebral metabolic rates of any element, normal (or therapeutically normalized) arteriovenous differences would reflect adequate match between blood flow and cerebral extraction or production of the studied elements. This interpretation of cerebral hemometabolism has significant implications in the understanding and, eventually, management of cerebral hemometabolism in acute intracranial disorders.

From the above, it becomes apparent that monitoring and management of cerebral perfusion pressure alone is somewhat "empirical", without any metabolic information. It is also apparent that monitoring and management of cerebral blood flow is also somewhat "empirical", without metabolic information. Therefore, the closest modality of monitoring of cerebral hemometabolism is that which involves monitoring of arteriovenous differences. For example, for any value of cerebral metabolic rate of oxygen consumption, cerebral blood flow will be adequate if the arteriovenous difference

is spontaneously (or therapeutically) in the normal range. This interpretation of cerebral hemometabolic physiology strongly agrees with the Fick principle, and the metabolic control of cerebral blood flow, thus constituting the basis for jugular oxygenation monitoring and management. The latter was introduced in the mid 1980's[2, 3] and has rapidly been expanded[1, 4-8, 11, 13, 14, 18, 22, 25].

Global cerebral oxygenation is easy to monitor, either by frequent intermittent[3, 9] or continuous[1-8, 11, 13, 18, 22, 25] arterial and jugular bulb measurements. Regional cerebral oxygen extraction is also measurable, by intermittent assessment with positron emission tomography[19, 27]. Relevant limitations of positron emission tomography are still due to its technical features, particularly in regard to unstable patients with acute intracranial disorders and severe intracranial hypertension. Focal (not regional) near-infrared cerebral oxygenation monitoring is now available, on a preliminary basis[20]. Unfortunately, relevant limitations are still a concern before this technique comes into routine clinical practice, as described elsewhere[10].

Methodology

Arteriojugular Monitoring

Intermittent jugular monitoring (with any clinically acceptable catheter) is unlikely to result in clinically relevant complications. However, it requires relatively frequent blood sampling (under most circumstances every 4–6 hours). Continuous fiberoptic jugular oxygenation monitoring requires adequate nursing and placement of the catheter, with its tip approximately 0.5 cm away from the superior wall of the jugular bulb. It also requires strict positioning and repositioning of the patient's head, neck, and shoulders, in order to avoid artifactual recordings. A major advantage of jugular bulb catheterization is that it does not require lowering and/or turning the head, with catheter insertion at the level of the cricoid cartilage. The basic guide for adequate catheter placement is a resistance which will be felt when the tip of the catheter reaches the superior wall of the jugular bulb. This will indicate that the catheter should be mobilized outwards by no more than 0.5 cm, to avoid endotelial damage and artifactual recordings.

Alternative Techniques to Measure Cerebral Oxygenation

Regional cerebral oxygenation by positron emission tomography, unlike jugular catheterization, cannot be done at bedside. In addition, even assuming that the patient will be kept on a steady state condition during transport, the procedure requires placing the head lower than in the desirable position for patients with acute intracranial hypertension. It is usually a long procedure, and the positron emission tomography rooms are usually not as well equiped as the neuro-intensive care rooms, regarding monitoring and management of elevated intracranial pressure. Perhaps even more limiting is the fact that a single test is very costly, and a repeat test, e.g. when increased regional cerebral oxygen extraction is found and managed, would add substantially to the time the patient is "unprotected", away from the intensive care environment, besides adding substantially to the cost of medical care. Perhaps, in the future, a bedside, and less expensive technique for positron emission tomography will contribute significantly to the intermittent monitoring and eventually management of increased regional cerebral oxygen extraction (oligemic cerebral hypoxia). This would be important, because oligemic cerebral hypoxia appears to represent a preischemic event.

Near-infrared spectroscopy appears to be a promising technique for monitoring cerebral oxygenation because it is noninvasive. However, its purely focal feature and its insufficient resolution to clearly detect veno-capillary oxygenation exclusively make the technique limited in the routine monitoring of acute intracranial disorders. Furthermore, even if the resolution of near-infrared spectroscopy is improved, a truly multiregional monitoring system (with multiple probes) would be required, in order to allow selective optimization of multiregional cerebral oxygenation. Such a multiregional monitoring system would appear to become of clinical relevance, should it become available in the future.

Results and Discussion

Clinical Experience With Jugular Monitoring

Isolated, observational arteriojugular measurements have been reported since 1942[15]. However, arteriojugular monitoring and management was first adopted by our team in 1981. We have now experienced frequent intermittent or continuous arteriojugular oxygen monitoring in over 600 acutely comatose patients, with gratifying results[2-11, 13]. Initially, we were able to identify and manage a condition of abnormally decreased cerebrovenous oxygenation in the presence of normal arterial oxygenation. This condition was therefore termed *oligemic cerebral hypoxia*[2, 3, 5, 6, 9]. At the same time, we proposed a new variable of global cerebral oxygenation, which was termed *cerebral extraction of oxygen* (CEO_2)[4-11]. The CEO_2 is simply the difference between arterial and jugular bulb oxyhemoglobin saturation. The CEO_2 was proposed to replace the arteriojugular difference in oxygen content ($AVDO_2$), because the latter was found to mistakenly suggest global cerebral hyperperfusion relative to oxygen consumption (cerebral hyperemia) in acute anemia, when the jugular oxygen saturation, oxygen content, and oxygen tension were all abnormally low, in good agreement. Thus, the CEO_2 independently informed about inadequate global cerebral oxygenation, for being a more stable calculation (not influenced by low hemoglobin values, such as the calculated $AVDO_2$)[12].

The most recent outgrowth of our methodology and technique is a new concept of global cerebral hemometabolism, which was termed *cerebral hemodynamic reserve* (CHR)[6, 8]. The CHR was born as an attempt to integrate cerebral hemodynamics and metabolism with intracranial compliance. This is because intracranial compliance measurements provide no information on cerebral hemodynamics and metabolism. Furthermore, cerebral hemodynamic and metabolic measurements alone provide no information about intracranial "tightness" (decreased compliance). The physiologic expression of the CHR is that it allows quantification of the global cerebral microcirculatory tolerance to decreases in cerebral perfusion pressure, unlike any previously reported variable or concept of cerebral hemometabolism.

Cerebral hemodynamic reserve is simply the ratio of percent changes in global cerebral oxygen extraction (the ratio of oxygen consumption to blood flow), to percent changes in cerebral perfusion pressure, under circumstances of decreases in cerebral perfusion pressure. In the two preliminary reports[6, 8], decreased cerebral perfusion pressure was assessed as a consequence of increased intracranial pressure. We found that, under circumstances of increased intracranial "tightness", as estimated by computerized tomography of the head, the CHR tended to become worse (compromised)[8]. This was particularly true when the intracranial pressure was only mildly increased. Thus, we believe that the CHR may herald a new era of monitoring and management of acute, predominantly diffuse intracranial disorders, allowing a fine tuned understanding of cerebral hemometabolic changes. This is likely to explain previously unknown causes of neurologic deterioration, at the same time when novel management strategies can be evaluated in a most objective fashion. An accurate interpretation of either intermittent or continuous assessment of the CHR can be found in a recent report[8].

Multidisciplinary Applications

Several conditions may be associated with acutely increased intracranial "tightness". Among those are acute brain trauma, hepatic metabolic encephalopathy, some types of meningitis and meningoencephalitis, spontaneous intracranial hemorrhage, and perhaps septic encephalopathy, among others. In addition, jugular oxygenation monitoring has been experienced successfully in the operating room in recent years[1, 18, 22]. Thus, the potential for cerebral oxygenation monitoring is rather broad, even beyond neurosurgical and neurological intensive care, and we do believe that cerebral oxygenation monitoring along with other monitoring techniques will definitively herald a new era in the comprehensive management of these acute intracranial disorders and anesthesia.

Cerebral Hypoxia

In the normal animal brain, global cerebral hypoxia has been shown to be virtually irrelevant, so that cerebral venous oxygen content levels as low as 2 vol% only have been proposed as safety parameters for therapies aimed at decreasing cerebral blood flow[26], when the normal jugular oxygen content is in the range of 12.5 vol%[15]. In the severely injured human brain, however, jugular oxyhemoglobin saturation values in the range of 30% have been reported as being associated with profound neurologic deterioration[4, 13]. Under most clinical conditions, such levels of jugular oxyhemoglobin saturation are associated with jugular oxygen content values of 3.5 vol% or higher, depending on the magnitude of acute anemia in the neuro intensive care environment. Therefore, threshold values of cerebrovenous oxygenation for neurologic deterioration in already comatose patients are higher than those of brain energy failure in the non-injured animal brain[26]. Thus, caution should always be exercised in extrapolating findings from normal animal physiology to that of the severely injured human brain.

Conclusions

From the above, it appears that arteriojugular monitoring will play a major role in the future, not only because arteriojugular oxygen measurements inversely correlate well with mean regional cerebral blood flow values[23], but also because from the same catheter a variety of additional measurements (such as glucose and lactate concentrations) can be performed. Particularly important for the experienced neuro-intensivist is the fact that arteriojugular monitoring is a bedside technique, unlikely to jeopardize patient care.

References

1. Croughwell ND, Prasco P, Blumenthal JA, Leone BJ, White WD, Reves JG (1992) Warming during cardiopulmonary bypass is associated with jugular bulb desaturation. Ann Thorac Surg 53: 827–832

2. Cruz J, Allen SJ, Miner ME (1985) Hypoxic insults in acute brain injury. Crit Care Med 13: 284 (Abstract)

3. Cruz J, Miner ME (1986) Modulating cerebral oxygen delivery and extraction in acute traumatic coma. In: Miner ME, Wagner KA (eds) Neurotrauma. Treatment, rehabilitation, and related issues, ed 1. Butterworths, Boston, pp 55–72

4. Cruz J (1988) Continuous versus serial global cerebral hemometabolic monitoring: applications in acute brain trauma. Acta Neurochir [Suppl] 42: 35–39

5. Cruz J, Miner ME, Allen SJ, Alves WM, Gennarelli TA (1990) Continuous monitoring of cerebral oxygenation in acute brain injury: injection of mannitol during hyperventilation. J Neurosurg 73: 725–730

6. Cruz J, Miner ME, Allen SJ, Alves WM, Gennarelli TA (1991) Continuous monitoring of cerebral oxygenation in acute brain injury: assessment of cerebral hemodynamic reserve. Neurosurgery 29: 743–749

7. Cruz J, Gennarelli TA, Alves WM (1992) Continuous monitoring of cerebral oxygenation in acute brain injury: multivariate assessment of severe intracranial "plateau" wave. Case report. J Trauma 32: 401–403

8. Cruz J, Gennarelli TA, Alves WM (1992) Continuous monitoring of cerebral hemodynamic reserve in acute brain injury: relationship to changes in brain swelling. J Trauma 32: 629–635

9. Cruz J, Gennarelli TA, Hoffstad OJ (1992) Lack of relevance of the Bohr effect in optimally ventilated patients with acute brain trauma. J Trauma 33: 304–311

10. Cruz J, Raps EC, Hoffstad OJ, Jaggi JL, Gennarelli TA (1993) Cerebral oxygenation monitoring. Crit Care Med (in press)

11. Cruz J (1993) Combined continuous monitoring of systemic and cerebral oxygenation in acute brain injury: preliminary observations. Crit Care Med (in press)

12. Cruz J, Jaggi JL, Hoffstad OJ (1993) Cerebral blood flow and oxygen consumption in acute brain injury with acute anemia: an alternative for the CMRO$_2$? Crit Care Med (in press)

13. Cruz J (1993) On-line monitoring of global cerebral hypoxia in acute brain injury. Relationship to intracranial hypertension. J Neurosurg (in press)

14. Garlick R, Bihari D (1987) The use of intermittent and continuous recordings of jugular venous bulb oxygen saturation in the unconscious patient. Scand J Clin Lab Invest [Suppl 188] 47: 47–52

15. Gibbs EL, Lennox WG, Nims LF, Gibbs FA (1942) Arterial and cerebral venous blood. Arterial-venous differences in man. J Biol Chem 144: 325–332

16. Guillaume J, Janny P (1951) Manometrie intracrânienne continue: intérêt de la méthod et premier résultats. Rev Neurol 84: 131–142

17. Kety SS, Schmidt CF (1948) The nitrous oxide method for the quantitative determination of cerebral blood flow in man: theory, procedure, and normal values. J Clin Invest 27: 476–483

18. Kubawara M, Nakajima N, Yamamoto F, Fujita T (1992) Continuous monitoring of blood oxygen saturation of internal jugular vein as a useful indicator for selective cerebral perfusion during aortic arch replacement. J Thorac Cardiovasc Surg 103: 355–362

19. Lenzi GL, Frackowiack RS, Jones T (1982) Cerebral oxygen metabolism and blood flow in human cerebral ischemic infarction. J Cereb Blod Flow Metab 2: 321–335

20. McCormick P, Stewart M, Goetting M, Dujovny M (1991) Noninvasive cerebral optical spectroscopy for monitoring cerebral oxygen delivery and hemodynamics. Crit Care Med 19: 89–97

21. Miller JD (1985) Head injury and brain ischemia. Implications for therapy. Br J Anesth 57: 120–130
22. Nakajima T, Masakasu K, Hayashi Y, Kitaguchi K, Uchida O, Takaki O (1992) Clinical evaluation of cerebral oxygen balance during cardiopulmonary bypass: on-line continuous monitoring of jugular venous oxyhemoglobin saturation. Anesth Analg 74: 630–635
23. Obrist WD, Langfitt TW, Jaggi JL, Cruz J, Gennarelli TA (1984) Cerebral blood flow and metabolism in comatose patients with acute head injury. Relationship to intracranial hypertension. J Neurosurg 61: 241–253
24. Rosner MJ, Daughton S (1990) Cerebral perfusion pressure management in head injury. J Trauma 30: 933–941
25. Sheinberg M, Kanter MJ, Robertson CS, Contant CF, Narayan RK, Grossman RG (1992) Continuous monitoring of jugular venous oxygen saturation in head-injured patients. J Neurosurg 76: 212–217
26. Sutton LN, McLaughlin AC, Dante S, Kotapka M, Sinwell T, Mills E (1990) Cerebral venous oxygen content as a measure of brain energy metabolism with increased intracranial pressure and hyperventilation. J Neurosurg 73: 927–932
27. Ter-Pogossian MM, Eichling JO, Davis DO, Welch MJ (1970) The measure in vivo of regional cerebral oxygen utilization by means of oxyhemoglobin labeled with radioactive oxygen-15. J Clin Invest 49: 381–391

Correspondence and Reprints: J. Cruz, M. D., Division of Neurosurgery, Hospital of the University of Pennsylvania, 3400 Spruce Street, Philadelphia, PA 19104, U.S.A.

Acta Neurochir (1993) [Suppl] 59: 91–97

Technical Considerations in Continuous Jugular Venous Oxygen Saturation Measurement

N.M. Dearden and **S. Midgley**

Department of Clinical Neurosciences, Western General Hospital, Edinburgh, U.K.

Summary

Fibreoptic reflection oximetry allows continuous in-vivo estimation of jugular venous oxygen saturation. In combination with pulse oximetry the oxygen extraction ratio $SaO_2 - SjO_2/SaO_2$ can be derived enabling identification of states of global luxury perfusion, normal coupling of global cerebral blood flow with global cerebral metabolism, global cerebral hypoperfusion and global cerebral ischaemia. Several technical difficulties may arise affecting the accuracy of SjO_2 recordings which must be recognised by the clinician before medical intervention is contemplated.

Keywords: Cerebral ischaemia; jugular bulb oxygen saturation; reflection oximetry.

Introduction

Ischaemic brain damage is commonly found at autopsy following severe cerebral trauma[7, 8] and also contributes to morbidity in patients who survive[5]. With the introduction of pulse oximetry greater attention has been addressed to monitoring and maintaining adequate levels of arterial oxygen saturation both during resuscitation and intensive care of neurosurgical patients. Until recently, however, most centres have relied upon continuous monitoring of intracranial pressure (ICP) and blood pressure (BP) and therapeutic maintenance of ICP below 20 mm Hg and CPP above an arbitrary limit 60 mm Hg in an attempt to maintain adequate cerebral perfusion and oxygen delivery[11]. The merits of this approach are being increasingly questioned since neurosurgical patients may have vascular distortion from intracranial shifts, deranged autoregulation, and cerebral vasospasm in association with intracranial hypertension, and may require considerably higher CPP to avert cerebral ischaemia[2].

During the last decade several groups have attempted to identify patients at risk from cerebral ischaemia by intermittent measurements of cerebral blood flow (CBF) and cerebral metabolism of oxygen and lactate (CMRO$_2$, CMRL)[13]. The relationship between global CMRO$_2$ and CBF can be described according to the Fick principle from measurement of the arterial-jugular venous oxygen content difference (AJDO$_2$). Thus AJDO$_2$ = CMRO$_2$/CBF. Since AJDO$_2$ = haemoglobin concentration $\times 1.34 \times$ arterio-venous oxygen saturation difference/100, if arterial oxygen saturation, haemoglobin concentration, and the position of the haemoglobin dissociation curve remain constant, the ratio of global CBF/CMRO$_2$ is proportional to the jugular bulb venous oxygen saturation (SjO$_2$). Using pulse oximetry or intermittent in vitro estimations of arterial oxygen saturation (SaO$_2$) measurements of global cerebral oxygen extraction ratio (OER = SaO$_2$ − SjO$_2$/SaO$_2$) can also be derived[3]. In the absence of anaemia or a sudden increase in FIO$_2$, increases in SjO$_2$ over 75% (OER $\leqslant$ 22%) suggest "luxury perfusion" (although areas of regional ischaemia or infarction may be present) and reductions of SjO$_2$ below 54% (OER $\geqslant$ 43%) indicate relative hypoperfusion while below 40% (OER $\geqslant$ 57%) global cerebral ischaemia is likely and will be associated with an increased production of lactic acid and a raised lactate oxygen index LOI (LOI = −AJD/AJDO$_2$) {AJDL (arterial-jugular venous lactate content difference) = CMRL/CBF}[13, 14].

A major limitation of intermittent measurements remains the ability to only monitor the patient at a particular moment in time. Furthermore, during periods of low cerebral blood flow sample contamination with extracerebral venous blood is more likely. Recently it has become possible, using fibreoptic technology, to measure SjO_2 continuously.

Methodology

Current Fibreoptic Technology for Continuous Oximetry

In-vivo oximetry has been possible for over 30 years although only recently have accurate, practical and reliable systems emerged for clinical use[4, 10].

Since 2 wavelength pulmonary flotation catheter systems proved unreliable, a 3 wavelength system was introduced by Abbott Critical Care Systems, Chicago, U.S.A. This Oximetrix 3 system comprises a processor, optical module and fibreoptic catheter. Light of red and near infra red wavelengths is transmitted at 1 ms intervals down a transmitting fibre and reflected along a receiving fibre to a photoelectric sensor. Reflected signals are averaged over 5 seconds and updated every second. The system analyses raw optical data, uses a patented Digital Signal Filter to reject vessel wall artefact and displays the trended and current oxygen saturation. Reflected light intensity is also displayed as a bar between 2 dotted lines. High reflected light intensity indicates vessel wall artefact while low intensity suggests catheter obstruction. Inability to display a light intensity signal larger than a dot between the upper and lower dotted lines signifies damage to the fibreoptics.

By using 3 wavelengths two independent reflected light intensity ratios R1 and R2 are calculated. Oxygen saturation can now be derived from 2 formulae; $SO_2 = A1 + B1*R1 + C1*[Hb] + D1*R1*[Hb]$ and $SO_2 = A2 + B2*R2 + C2*[Hb] + D2*R2*[Hb]$. (A1,B1,C1,D1,A2,B2,C2, and D2 are constants). The 2 equations can be solved algebraeically eliminating the need for user input of [Hb]. Thus $SO_2 = ((A1*C1-A2*C1) + (B1*C2-A2*D2)*R1 + (A1*D2-B2*C1)*R2 + (B1*D2-B2*D1)*R1*R2)/C2-C1 - D1*R1 + D2*R2$. This reduces the sensitivity of the instrument to changes in haematocrit although other patent protected formulae are necessary to account for changes in pH and blood flow velocity. The use of 2 reflected light intensity ratios also virtually eliminates any contribution from carboxyhaemoglobin to the computed saturation. The catheter can be calibrated in 2 ways.

Firstly an in-vitro calibration is done before insertion (pre-insertion calibration). This is achieved using a standard optical reference rather than a blood sample and is possible because of a patent controlled standard design of the optical spacing of the monofilament optical fibres. This calibration was designed for pulmonary artery flotation catheters and by using a controlled colour reference significantly narrower confidence intervals of mixed venous saturation for a given series of reflected light intensity ratios are achieved while errors associated with blood sampling and in vitro measurements are eliminated. The pre-insertion calibration is also useful as it provides a check of the integrity of the whole system prior to catheter insertion.

Secondly the catheter may be calibrated in-vivo by comparing and adjusting catheter recordings with values obtained by analysis of an aspirated sample of jugular bulb venous blood in a laboratory Co-Oximeter. This should be done either if pre-insertion calibration was omitted prior to insertion or after the catheter has been left in place for some time since clinical experience with fibreoptic pulmonary flotation catheters indicates significant drift

from laboratory in-vitro measured blood saturation 12–36 hours after insertion. Following in-vivo calibration recordings change with fluctuations in pH and haematocrit and the confidence interval of saturation for a given set of light intensity ratios widens. In-vivo calibration is compromised if blood sampling errors occur, if calculated (from blood gas tensions) rather than actual measurements of blood oxygen saturations are used or if the laboratory Co-Oximeter is not correctly calibrated (Fig. 1). Furthermore, a small error is incurred if the laboratory Co-Oximeter includes carboxyhaemoglobin and methaeoglobin in the total haemoglobin amount. In this event SO_2 (Oximetrix) $= (SO_2(laboratory)/1 - ((HbCO\% + Hb Met\%)/100)$.

The Oximetrix 3 System for Measurement of Jugular Venous Oxygen Saturation

Two fibreoptic catheters, originally designed for SaO_2 measurements in neonates, have been used for measurement of jugular venous oxygen saturation, one 25 cms and the other 40 cms in length. The latter has been evaluated clinically by our group and the recordings obtained compared with in-vitro measurements using an Instrumentation Laboratories 282 Co-Oximeter[1]. Following pre-insertion calibration there was statistically significant but clinically poor correlation between the 2 systems and the catheter system usually overread. It was necessary to perform an in-vivo calibration and repeat this every 12 hours in order to maintain the limits of agreement (LA = Mean difference $\pm 1.96*SD$ of difference) below 5%. Since the publication of our study we have evaluated a further twenty-eight 40 cm Shaw opticaths and compared them with a new stiffer 25 cm catheter specially designed for use in the jugular bulb since the manufacturers were of the opinion that less vessel wall artefact might be afforded by a stiffer catheter. We confirm our previous findings that following pre-insertion calibration the 40 cm catheter overreads by 7% with wide limits of agreement (−7.3 to +21.43%). Correlation regression analysis of 40 cm catheter readings after pre-insertion calibration with in-vitro measurements from an Instruments Laboratories 482 Co-Oximeter was significant $r = 0.78$ (Catheter $= 6.95 + 1*Co$-Oximeter) (M. S. Dearden N.M. and Miller J.D., unpublished observations). We found the new style catheter to be less reliable clinically with a greater tendency to overread after pre-insertion calibration, wider limits of agreement and poorer performance after in-vivo calibration. Possible explanations for the discrepancy between the performance of pulmonary artery flotation catheters and jugular vein catheters after pre-insertion calibration include different catheter materials and thickness, and the difference in direction of flow and the flow characteristics of the blood. It appears that a new catheter specifically designed for use in the jugular bulb possibly employing a "J Loop" will be required to improve the reliability and accuracy of continuous jugular venous oximetry.

The scheme for insertion and calibration of the oximetrix 3 system using a Shaw 40 cm fibreoptic catheter based on our findings is given in Table 1.

Considerations Relevant to Positioning of the Catheter

The jugular bulb is a dilatation of the rostral internal jugular vein just below the jugular foramen. In normal patients 80–90% of blood drains from both cerebral hemispheres via intracranial venous sinuses through the sigmoid sinuses and onward into the right internal jugular vein[17], although simultaneous bilateral jugular bulb saturations are equal and probably represent drainage from all parts of the brain[6, 15]. Free venous crossflow through the transverse sinus is present in about 75% of normal children. In the

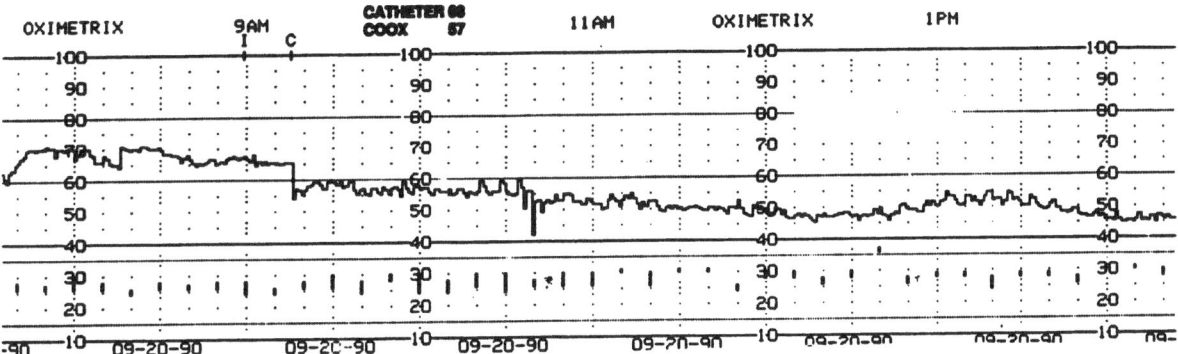

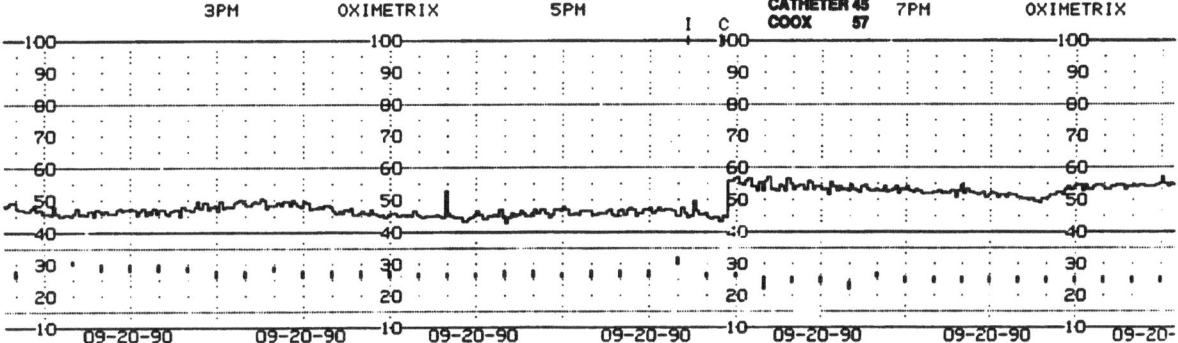

Fig. 1. Error in SjO$_2$ recording following in-vivo calibration of a fibreoptic catheter. At 9 a.m. the catheter is reading 68% but an aspirated blood sample measures 57% on the Co-Oximeter. The catheter is adjusted to correspond to the latter value. When subsequently the Co-Oximeter is calibrated it is found to be under-reading by 12% and the fault is corrected. At 5.40 p.m. in-vivo calibration is repeated and demonstrates the catheter underreading by 12%. Between the 2 in-vivo calibrations the catheter system was therefore under-estimating SjO$_2$ because a poorly calibrated Co-Oximeter was used to analyse jugular venous blood samples

Table 1. *Technique for Insertion of a 40 cm Shaw Opticath for SJO$_2$ Monitoring with the Oximetrix 3 System*

1. Pre-Insertion Calibration of Opticath
2. Seldinger technique to insert introducer
3. Attach Opticath to heparin/saline flush and prime
4. Advance Opticath until tip in jugular bulb
5. Withdraw introducer over Opticath
6. Confirm free aspiration of blood via Opticath
7. Verify satisfactory light intensity
8. Lateral cervical X ray to verify position
9. In-vivo calibration, aspirate blood slowly
10. Repeat in-vivo calibration 12 hourly

presence of focal intracranial pathology patterns of drainage may change so that differences in the SjO$_2$ are occasionally evident between right and left[9, 16]. Some investigators therefore choose to measure SjO$_2$ from the internal jugular vein on the side of focal pathology[14]. Our policy, however, is to sequentially manually compress the internal jugular veins and observe the effects on ICP selecting the side of greater rise in ICP as representative of predominant drainage. In the event of equal rises, which may indicate free communication across the transverse sinus, the right side is chosen because it is usually easier to cannulate.

The internal jugular vein descends adjacent to the lateral aspect of the internal and common carotid arteries together with the vagus nerve and enclosed with them in the carotid sheath. It terminates behind the medial part of the clavicle by joining the subclavian vein to form the brachiocephalic vein. Inferior petrosal sinus, mastoid emissary veins, pharyngeal plexus, facial vein, lingual vein, superior and middle thyroid veins and the jugular lymph trunk join the internal jugular vein throughout its passage in the neck. For accurate measurement of cerebral venous oxygen saturation it is therefore essential to confirm the position of the catheter tip high in the jugular bulb, preferably above the upper border of the second cervical vertebra on lateral cervical X ray, at regular intervals. We cannulate the internal jugular vein, with the patient maintained in a head up position whenever possible. We note the ICP, then turn the head slightly away from the selected side, puncture the skin just lateral to the pulsation of the carotid artery at the level of the thyroid cartilage and slowly advance the needle, with intermittent aspiration, towards the ipsilateral external auditory meatus. The needle will be felt to puncture the carotid sheath anteriorly before entering the vein. Should the posterior sheath be encountered without aspiration of venous blood the needle should be carefully removed using intermittent aspiration as frequently the jugular vein has been entered while partially collapsed and free aspiration of blood occurs during needle withdrawl. The vein is then cannulated using a Seldinger technique. For insertion of a 40 cm 4F gauge fibreoptic catheter we use a 115.20 14 gauge Vygon Leadercath as an introducer. The fibreoptic catheter is advanced to between 11 and 14 cms from the skin, when the tip may be felt to about the base of the skull, the introducer is carefully withdrawn over the fibreoptic catheter and the system secured with sutures and sterile dressings. After insertion the head is turned back into neutral position and the ICP observed. We consider a sustained rise in ICP of more than 5 mm Hg at this time over the pre-insertion value an indication to remove the catheter. Finally the catheter tip position is confirmed radiographically. Our contraindications to the technique include bleeding diathesis, local infection, local neck trauma and any impairment to cerebral venous drainage. Provided these contra-indications are observed and the system is continuously flushed with 3 mls per hour of 2 units per ml heparinised saline, impaired cerebral venous drainage is negligible and the risk of significant venous thrombosis appears less than 5%. If for any reason blood cannot be freely aspirated from the catheter system, it is our policy to remove the catheter immediately.

Results and Discussion

Problems Encountered During Clinical Use of Continuous SjO_2 Monitoring

With the emergence of fibreoptic technology it has become possible to continuously monitor SjO_2 and in combination with pulse oximetry to display immediately changes in the ratio of global CBF to $CMRO_2$. However, the clinician must be aware of the limitations of any system they use to measure jugular venous oxygen saturation in order to correctly interpret recordings. In addition to the problems associated with calibration outlined above, we have encountered several difficulties when using the Shaw 40 cm opticath with the Oximetrix 3 system, examples of which are described below.

Poor Light Intensity

Light intensity should always be displayed during measurement of SjO_2. This appears as a white bar (narrower than seen with pulmonary vascular catheters) lying between an upper and lower dotted line (Fig. 3b). If the fibreoptic catheter becomes obstructed a low light intensity display may be encountered with a dot lying at the bottom of the light intensity display. Saturation recordings should be considered unacceptable even when a light intensity band crosses the lower dotted line of the light intensity display. The catheter should be aspirated until blood may be freely sampled and a normal light intensity is displayed and then flushed with heparinised saline, otherwise the catheter should be replaced. In contrast a broad light intensity bar which crosses the upper boundary line of the display does not necessarily indicate inaccurate satuation data but a dot above this line signifies the catheter tip is up against the vessel wall and that the reading (characteristically between 85–95%) is invalid. This problem is frequently encountered when turning a

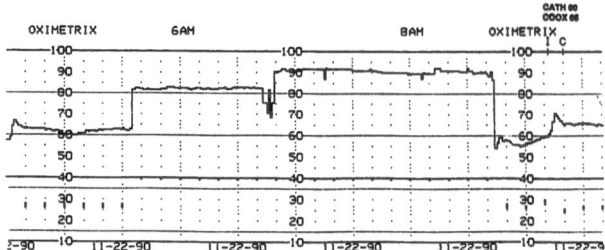

Fig. 2. Trace of normal SjO_2 with satisfactory reflected light intensity display until 5.45 a.m. when SjO_2 suddenly increases in association with high light intensity suggesting the catheter tip is against the vessel wall. After 8.20 a.m. SjO_2 abruptly declines with return of acceptable light intensity readings. In-vivo calibration at 8.40 a.m. confirms concordance within 5% of a sample analysed in the laboratory Co-Oximeter

patient's head or during other nursing procedures and is usually rectified by adjusting head position although occasionally the catheter will need repositioning (Fig. 2). The error caused by vessel wall artefact depends on the relative amounts of signal returning from the vessel wall and the blood.

Sampling and Calibration Errors

In addition to sample handling techniques, in-vivo calibration errors of the Oximetrix 3 system may arise if significant changes in saturation are displayed during aspiration of a blood sample. The risks of this are minimised by slow aspiration (1 ml of blood over 1 minute) to avoid contamination of the sample with extracerebral venous blood. Figure 3a shows a catheter recording SjO_2 around 70%. Despite slow aspiration the value falls on 2 separate occasions by 10%. There is no apparent change in light intensity to suggest cathether artefact and as the Co-Oximeter reads 59% sample contamination could be suspected. However, after in-vivo calibration the catheter and Co-Oximeter are rechecked and found to corroborate and yet a dip of 10% without change in light intensity still occurs with slow sampling from the catheter. On the following day when in-vivo calibration was repeated however, catheter and Co-Oximeter readings were similar and no fall in SjO_2 was observed during sample aspiration (Fig. 3b). These observations are hard to explain but sample contamination appears unlikely rather it appears the catheter was initially overreading. Dips in saturation during venous sampling may therefore represent an idiosyncrasy of

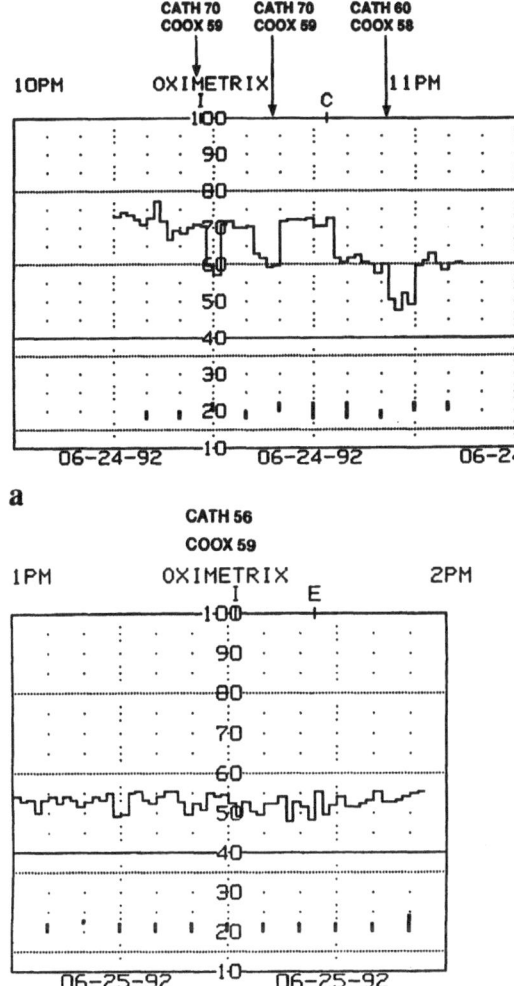

a

b

Fig. 3. (a) Unexplained effect observed during slow aspiration (1 ml over 120 seconds) of jugular bulb blood through a fibreoptic catheter recording a normal SjO₂ with a good reflected light intensity display. On 2 separate occasions the catheter reads 70% prior to aspiration but falls below 60% without change in reflected light intensity as blood is withdrawn. Sample saturation is 59% when measured on the laboratory Co-Oximeter. In-vivo calibration is performed and the catheter adjusted to correspond to the Co-Oximeter reading of 59% at 10.45 a.m. A further sample taken at 10.55 a.m. confirms close corroboration (within 2%) of catheter and Co-Oximeter but a fall of 10% occurs during sample aspiration again without change in reflected light intensity. (b) Recording from the same catheter the following day. SjO₂ lies at the lower end of the normal range and reflected light intensity is satisfactory with a bar lying between the 2 dotted lines. At 1.30 p.m. a sample drawn through the fibreoptic catheter produces no dip in saturation or light intensity display. Catheter and Co-Oximeter readings are within 3% indicating preservation of accuracy since the previous in-vivo calibration

the catheter system, perhaps from changes in the flow characteristics of the blood or movement of the catheter tip during aspiration, that requires further investigation.

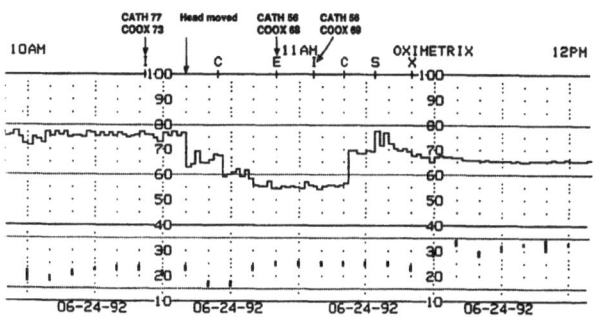

Fig. 4. Trace demonstrating the effect of head movement on SjO₂ recording. Prior to head movement in-vivo calibration shows catheter and Co-Oximeter readings are within 5% (*I*). Head turning during removal of a pillow causes a sudden fall in displayed SjO₂ without significant change in reflected light intensity at 10.35 a.m. At *C* the catheter is adjusted to correspond to the Co-Oximeter reading leading to a step fall of 4%. At *E* a jugular venous blood sample is taken and analysed and shows the catheter to be underreading by 12%. A second in-vivo calibration is performed at 11.03 a.m. (*I*) demonstrating the catheter is underreading by 13%. This corresponds to the level of fall in SjO₂ recorded when head position was adjusted. The Catheter is recalibrated to correspond to the Co-Oximeter at 11.10 a.m. (*C*). *S* and *X* represent suction and X ray respectively. Reflected light intensity display remains satisfactory throughout

Effects of Head Movement

When used for SjO₂ measurement the Oximetrix 3 system is very sensitive to head position and head movement. We frequently observe sudden changes in displayed SjO₂ (usually a decrease) with or without changes in light intensity when the head position is changed even in paralysed ventilated patients. An example of this is shown in Fig. 4 where catheter and Co-Oximeter initially correspond but the catheter underreads after the head is moved; light intensity display appears satisfactory throughout.

SjO₂ "Waves"

During intracranial pressure B waves the SjO₂ may be observed to increase and decrease rhythmically either with or between rises in ICP. An example is shown in Fig. 5a where each time CPP falls SjO₂ repeatedly decreases into the hypoperfusion range. However, we have observed waves in the SjO₂ (during periods of adequate reflected light intensity) when no ICP or blood pressure oscillations are evident (Fig. 5b). The cause is coiling of the catheter in the internal jugular vein. This can be verified radiologically and corrected by replacing the catheter.

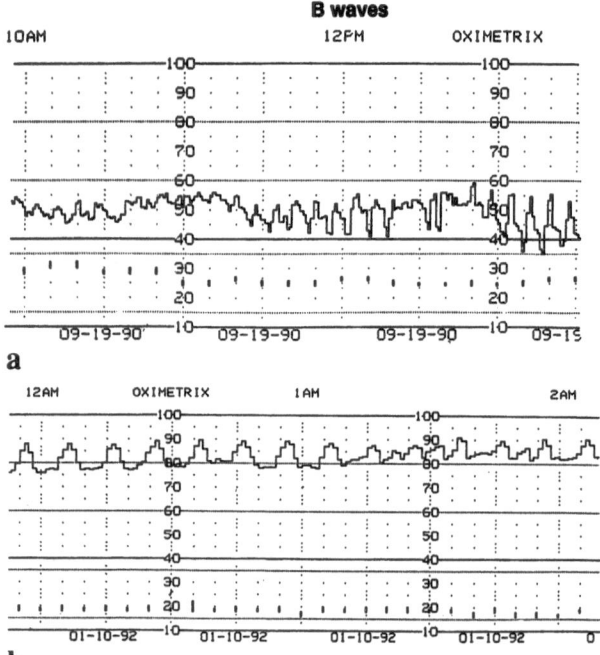

a

b

Fig. 5. (a) Recording of SjO_2 during intracranial pressure B waves. SjO_2 periodically falls below 50% when ICP rises and cerebral perfusion pressure falls. Reflected light intensity display is satisfactory. (b) Rhythmic fluctuations in SjO_2, documented when ICP and blood pressure were normal, due to coiling of the fibreoptic catheter inside the jugular vein. Note that reflected light intensity display remains acceptable

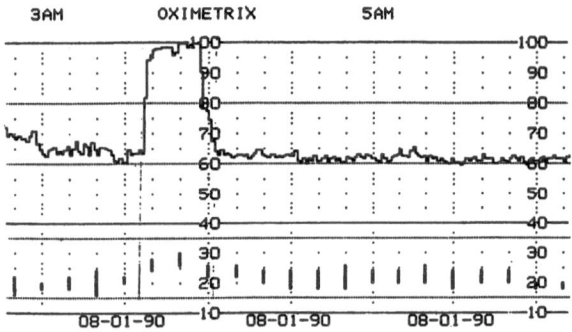

Fig. 6. Unexplained sudden rise in displayed SjO_2 from normal range to 100% just after 3.30 a.m. in a sedated paralysed undisturbed patient. SjO_2 falls abruptly to its previous level just before 4 a.m. No significant change in reflected light intensity display occurs

Sudden Changes in SjO_2

We have occasionally observed a sudden increase in SjO_2 up to 100% without change in light intensity during a period of clinical stability when a ventilated paralysed patient was completely undisturbed and FIO_2 remained unaltered (Fig. 6). The cause is unknown.

Conclusion

Continuous jugular venous oximetry is now clinically possible using the Oximetrix 3 fibreoptic system. Although there have been significant improvements in fibreoptic technology, considerable vigilance is required in preparing the system to obtain reliable clinical data and repeated verification of catheter position and calibration is necessary before clinical decisions are invoked from SjO_2 data obtained from the system. In the authors' opinion this is a useful technique in experienced hands but further work is required to design a more reliable system before recommending widespread introduction of continuous jugular venous oxygen saturation monitoring.

References

1. Andrews PJD, Dearden NM, Miller JD (1991) Jugular bulb cannulation: Description of a technique and validation of a new continuous monitor. Br J Anaesth 67: 553–558
2. Chan KH, Miller JD, Dearden NM, Andrews PJD, Midgley S (1992) The effect of changes in cerebral perfusion pressure upon middle cerebral artery blood flow velocity and jugular bulb venous oxygen saturation after severe brain injury. J Neurosurg 77: 55–61
3. Cruz J, Miner ME, Allen SJ, Alves WM, Gennarelli TA (1991) Continuous monitoring of cerebral oxygenation in acute brain injury: Assessment of Cerebral Hemodynamic reserve. Neurosurgery 129: 743–749
4. Enson Y, Briscoe WA, Polanyi ML, Cournand A (1962) In vivo studies with an intravascular and intracardiac reflection oximeter. J Appl Physiol 17: 552–558
5. Gentleman D, Jennett B (1990) Audit of transfer of unconscious patients to a neurosurgical unit. Lancet 335: 330–334
6. Gibbs EL, Lennox WG, Gibbs FA (1945) Bilateral internal jugular blood. Comparison of A-V differences, oxygen-dextrose ratios and respiratory quotients. Am J Psychiatry 102: 184–190
7. Graham DI, Adams JH, Doyle D (1978) Ischaemic brain damage in fatal non-missile head injuries. J Neurol Sci 39: 213–234
8. Graham DI, Ford I, Hume-Adams J, Doyle O, Teasdale GM, Lawrence AE, McLellan DR (1989) Ischaemic brain damage is still common in fatal non-missile head injury. J Neurol Neurosurg Psychiatry 52: 346–350
9. Lassen NA (1959) Cerebral blood flow and oxygen consumption in man. Physiol Rev 39: 183–238
10. Martin WE, Cheung PW, Johnson CC, Wong KC (1973) Continuous monitoring of mixed venous oxygen saturation in man. Anaesth Analg 52: 784–793
11. Miller JD (1985) Head injury and brain ischaemia-Implications for therapy. Br J Anaesth 57: 120–129
12. Miller JD (1991) Pathophysiology and management of head Injury. Neuropsychology 5: 235–261
13. Robertson CS, Grossman RG, Goodman JC, Narayan RK (1987) The predictive value of cerebral anaerobic metabolism with cerebral infarction after head injury. J Neurosurg 67: 361–368

14. Robertson CS, Narayan RK, Gokaslan ZL, Pahwa R, Grossman RG, Caram P, Allen E (1989) Cerebral arteriovenous oxygen difference as an estimate of cerebral blood flow in comatose patients. J Neurosurg 70: 222–230

15. Shenkin HA, Harmel MH, Ketty SS (1948) Dynamic anatomy of the cerebral circulation. Arch Neurol Psychiatry 60: 240–252

16. Shenkin HA, Spitz EB, and Grant FC (1948) Physiologic studies of arteriovenous anomalies of the brain. J Neurosurg 5: 165–172

17. Williams PL, Warwick R, Dyson M (eds) (1989) Grays Anatomy, ed 37. Churchill Livengstone, New York, pp 793–805

Correspondence and Reprints: Dr. N. Mark Dearden, MBChB, BSc, FFARCS, Honary Clinical Lecturer in Anaesthesia, Department of Anaesthesia, Leeds General Infirmary, Great George Street, Leeds LS1 3EX, U.K.

Acta Neurochir (1993) [Suppl] 59: 98–101

Desaturation Episodes after Severe Head Injury: Influence on Outcome

C. Robertson

Department of Neurosurgery, Baylor College of Medicine, Houston, Texas, U.S.A.

Summary

The relationship of jugular venous desaturation and neurological outcome was examined in 116 patients with severe head injury. Seventy-six episodes of jugular venous desaturation were prospectively identified in 46 (40%) of the patients. The etiology of the desaturations varied, including both systemic and cerebral causes. A poor neurological outcome was strongly associated with the occurrence of jugular venous desaturation.

Keywords: Jugular venous desaturation; severe head injury; outcome.

Introduction

Many factors contribute to the long-term neurological disability incurred by a patient who has sustained a severe head injury. The pathology produced by the injury, the patient's neurological status, and the patient's age have been identified as important determinants of outcome after head injury[7, 9, 11].

Other factors can occur during the early hospital course, which, if identified early, might prevent additional injury to the brain[8, 10]. Cerebral hypoxia/ischemia is one of the most important causes of secondary injury after trauma, and can occur from a variety of both systemic and cerebral causes, including intracranial hypertension, systemic hypotension, and hypoxia[5].

Continuous monitoring of jugular venous oxygen saturation has been proposed as a useful method for early detection of global cerebral ischemia due to these various mechanisms[1, 2–4, 6, 13]. Early detection and treatment of ischemia may prevent additional damage to the injured brain. The purpose of these studies was to study the relationship of jugular venous desaturation and neurological outcome.

Methods

Patient Characteristics

One hundred and sixteen patients have been studied. The average age was 31.5 ± 14.7 years. One-hundred and four (90%) were male. One hundred (86%) of the patients had closed head injuries and 16 (14%) were gunshot wounds. Ninety-two (80%) of the patients were in coma (GCS $\leq$ 8) on admission to the hospital; 24 (21%) of the patients had an initial GCS > 8 but deteriorated within the first 48 hours. The mean GCS was 6.9 ± 3.0 in the ER and 6.8 ± 2.3 on day 1 after injury.

Monitoring Protocol

$SjvO_2$ was monitored continuously using a fiberoptic oxygen saturation catheter placed in the jugular bulb. The correct position of the catheter was confirmed by x-ray. The calibration of the catheter was confirmed every 8 hours by measuring oxygen saturation in a blood sample drawn through the catheter. Episodes of jugular venous desaturation ($SjvO_2 < 50\%$) were investigated using a standard algorithm previously described[13].

Every attempt was made to start the $SjvO_2$ monitoring as soon after admission as possible. The $SjvO_2$ catheter was inserted an average of 17 ± 15 hours after admission to the NICU. Ninety-three (90%) were inserted within 24 hours of admission. The catheter was placed on the right side unless the left jugular venous circulation was demonstrated to be dominant. Ninety-five (82%) of the catheters were placed on the right side, and 20 (17%) were on the left. Patients were monitored for an average of 88 ± 76 hours.

Results

Reliability of the Continuous $SjvO_2$ Measurements

In previous studies[13], we had found the fiberoptic oxygen saturation catheter to be a reliable measure of $SjvO_2$ after the first in vivo calibration and when the light intensity value indicated good position of the catheter in the jugular bulb (n = 176, r = 0.87, p < 0.01). However, it can be difficult in some patients to maintain a good light intensity and the catheter can be positional in some patients. Therefore, it is important to verify the catheter $SjvO_2$ values by measuring SO_2 in a blood sample drawn through the catheter before making therapeutic decisions.

In the 116 patients that have been studied, 172 episodes of catheter desaturation were prospectively identified and investigated. On 95 (60%) of these episodes, SO_2 was found to be greater than 50% in a blood sample drawn through the catheter. In the remaining 76 episodes, the catheter reading was confirmed.

Incidence and Causes of Confirmed Jugular Desaturations

$SjvO_2$ was considered abnormally low if <50% for more than 10 minutes and if the catheter value was confirmed by measuring SO_2 in a blood sample drawn through the catheter. Seventy (60%) of the 116 patients had $SjvO_2$ values that never decreased below 50% during the period of monitoring. Seventy-six episodes of jugular desaturation were identified and confirmed by blood sampling in the remaining 46 (40%) of the 116 patients. Twenty-seven patients had one episode of desaturation, 11 patients had 2 episodes, 7 patients had 3 episodes, and 1 patients had 6 episodes.

Most of the episodes of desaturation were so short-lived that they could not have been identified with intermittent measurements. The episodes of desaturation lasted for an average time of 1.2 ± 1.6 hours (range 10 minutes–12 hours). Fifty-seven (75%) of the episodes lasted for <1 hour. The lowest recorded value for $SjvO_2$ during these episodes was between 40 and 50% in 42 of the episodes, between 30 and 40% in 15 of the episodes, and <30% on 12 occasions.

The episodes of desaturation were most common during the first 24 hours after admission. Twenty-eight (37%) of the episodes occurred on day 1, 13 (17%) on day 2, 10 (13%) on day 3, and 12 (16%) on day 4.

The causes of the episodes of jugular desaturation were both systemic and cerebral, including intracranial hypertension (n = 32), hypocarbia (n = 21), hypotension (n = 8), hypoxia (n = 6), anemia (n = 1), cerebral vasospasm (n = 1), and combinations of the above (n = 6).

Association of Jugular Desaturation and Neurological Outcome

Ninety-five patients were at least 3 months after injury at the time of this analysis. Thirty-three (35%) of the patients had a favorable outcome (good recovery or moderate disability). Thirty patients (32%) were severely disabled or vegetative and thirty-two (34%) were dead. As shown in Table 1, the incidence of jugular desaturation and long-term neurological outcome were strongly associated (p = 0.0007). Mortality rate was 18% in patients with no desaturations, compared to 46% in patients with one episode of desaturation and 71% in patients with multiple episodes of desaturations. The percentage of patients with a poor neurological outcome (severely disabled, vegetative, or dead) was 54% in patients with no desaturations, 77% in patients with one desaturation, and 88% in patients with multiple desaturations.

Several factors including age, severity of injury, and type of injury are known to be associated with outcome from severe head injury. We have previously reported that a reduced CBF is associated with a poor outcome. To examine the effect of these other factors on the relationship between neurological outcome and the incidence of jugular desaturation, logistic regression analysis was used.

Table 2 shows the potential confounding factors that were examined. The GCS on day 1 after injury and the type of injury (diffuse vs. mass lesion) had significant associations with outcome. The patients with episodes of desaturation were monitored earlier than the patients without desaturations. Since the desaturations were more common early after injury, it was possible that the start of monitoring might have confounded the relationship between the desaturations and outcome. Patients with a reduced CBF had a greater incidence of desaturations than patients

Table 1. *Relationship between the Number of Episodes of Jugular Desaturation and Neurological Outcome at 3 Months after Injury*

Number of Desaturations	3 Month Glasgow Outcome Score			Total Number
	GR/MD	SD/V	Dead	
None	26 (46%)	20 (36%)	10 (18%)	56
One	5 (23%)	7 (32%)	10 (46%)	22
Multiple	2 (12%)	3 (18%)	12 (71%)	17

GR/MD Good recovery/moderate disability; *SD/V* severely disabled/vegetative.

Table 2. *Potential Confounding Factors that were Examined for Relationship between Desaturations and Outcome*

	Number of Desaturations				3 Month Glasgow Outcome Score			
	None	One	Multiple	p Value*	GR/MD	SD/V	Dead	p Value*
Number	56	22	17		33	30	32	
Age	30.9 ± 2.1	27.7 ± 2.3	35.5 ± 4.3	0.29	28.1 ± 2.4	30.7 ± 2.8	34.2 ± 3.0	0.34
Sex (% male)	52 (92.9%)	18 (81.8%)	15 (88.2%)	0.35	31 (93.9%)	26 (86.7%)	28 (87.5%)	0.58
ER GCS	6.9 ± 0.4	6.6 ± 0.5	7.4 ± 0.9	0.73	7.7 ± 0.5	6.4 ± 0.4	6.6 ± 0.5	0.15
ER Pupils (% normal)	53 (94.6%)	19 (86.4%)	16 (94.1%)	0.78	33 (100%)	28 (90.3%)	29 (87.9%)	0.54
Day 1 GCS	7.0 ± 0.3	6.8 ± 0.5	6.4 ± 0.8	0.70	8.0 ± 0.4	6.8 ± 0.3	5.7 ± 0.4	0.0002
Day 1 Pupils (% normal)	54 (96.4%)	20 (90.9%)	17 (100%)	0.75	33 (100%)	29 (93.5%)	31 (93.9%)	0.89
Injury Type				0.89				0.02
Diffuse	17 (30.9%)	8 (36.4%)	5 (31.3%)		16 (50.0%)	8 (26.7%)	6 (19.4%)	
Mass	38 (69.1%)	14 (63.6%)	11 (68.8%)		16 (50.0%)	22 (73.3%)	25 (80.6%)	
Start of Monitoring								
ICP	14.1 ± 2.0	7.5 ± 0.9	7.9 ± 1.0	0.03	11.8 ± 1.9	13.6 ± 3.1	9.0 ± 1.1	0.32
SjvO$_2$	21.2 ± 2.5	13.5 ± 2.4	11.1 ± 1.5	0.03	20.0 ± 3.4	19.1 ± 3.1	13.7 ± 1.7	0.24
CBF	23.8 ± 2.5	14.9 ± 2.6	12.8 ± 1.8	0.01	23.1 ± 3.5	21.4 ± 3.1	14.8 ± 1.8	0.11
CBF Group				0.04				0.12
reduced	6 (10.7%)	7 (31.8%)	6 (35.3%)		2 (6.1%)	7 (23.3%)	10 (31.3%)	
normal	18 (32.1%)	2 (9.1%)	4 (23.5%)		11 (33.3%)	6 (20.0%)	7 (21.9%)	
elevated	32 (57.1%)	13 (59.1%)	7 (41.2%)		20 (60.6%)	17 (56.7%)	15 (46.9%)	

GR/MD Good recovery/moderate disability; *SD/V* Severely disabled/vegetative.

* p value is for analysis of variance when mean values were compared and Chi-square with categorical data.

Table 3. *Final Logistic Regression Model for Poor Neurological Outcome as Dependent Variable*

	Variable	Odds Ratio	p Value
Injury	Mass lesion	7.04	0.002
	Diffuse	1.00	
Day 1 GCS	3–8	4.07	0.035
	>8	1.00	
Pupillary Reactivity	one or both unreactive	4.24	0.017
	both reactive	1.00	
No. of Desaturations	≥ 2	13.77	0.009
	1	4.79	
	0	1.00	

with a normal or elevated CBF. All of the factors were included in the initial analysis.

The final best fit model, including only those factors which were found to be significant, is shown in Table 3. These data suggest that even when adjusted for all covariates, the occurrence of jugular desaturation was strongly associated with a poor neurological outcome.

Discussion

These studies indicate that episodes of jugular venous desaturation are common in patients with severe head injury, even when they are receiving intensive care with advanced cardiovascular and intracranial monitoring. These episodes of desaturation would not have been detected in most patients without continuous monitoring of the SjvO$_2$. Episodes of desaturation were more common in the first 24 hours after injury and in patients with a reduced CBF. The potential utility of these measurements is underscored by the observation that patients with confirmed episodes of oxygen desaturation had a higher mortality rate than those without such episodes. This association remained strong even when adjusted for covariates, including severity and type of injury.

References

1. Andrews, PJ, Dearden NM, Miller JD (1991) Jugular bulb cannulation: description of a cannulation technique and validation of a new continuous monitor. Br J Anaesth 67: 553–558
2. Cruz J (1988) Continuous versus serial global cerebral hemometabolic monitoring: applications in acute brain trauma. Acta Neurochir (Wien) [Suppl] 42: 35–39

3. Cruz J, Miner ME, Allen SJ, *et al* (1991) Continuous monitoring of cerebral oxygenation in acute brain injury: assessment of cerebral hemodynamic reserve. Neurosurgery 29: 743

4. Garlick R, Bihari D (1987) The use of intermittent and continuous recordings of jugular venous bulb oxygen saturation in the unconscious patient. Scand J Clin Lab Invest 47 [Suppl 188]: 47–52

5. Graham DI, Ford I, Adams JH, *et al* (1989) Ischaemic brain damage is still common in fatal non-missile head injury. J Neurol Neurosurg Psychiatry 52: 346–350

6. Hans P, Franssen C, Damas F, *et al* (1991) Continuous measurement of jugular venous bulb oxygen saturation in neurosurgical patients. Acta Anaesthesiol Belg 42: 213–218

7. Jennett B, Teasdale G, Braakman R, *et al* (1977) Prognosis in series of patients with severe head injury. Neurosurgery 4: 283

8. Kohi YM, Mendelow AD, Teasdale GM, *et al* (1984) Extracranial insults and outcome in patients with acute head injury – relationship to the Glasgow Coma Scale. Injury 16: 25–29

9. Miller JD, Becker DP, Ward JD, *et al* (1977) Significance of intracranial hypertension in severe head injury. J Neurosurg 47: 501–516

10. Miller JD (1985) Head injury and brain ischaemia-implications for therapy. Br J Anaesth 57: 120–129

11. Narayan RK, Greenberg RP, Miller JD, *et al* (1981) Improved confidence of outcome prediction in severe head injury: A comparative analysis of the clinical examination, multimodality evoked potentials, CT scanning and intracranial pressure. J Neurosurg 54: 751–762

12. Robertson CS, Contant CF, Gokaslan ZL, Narayan RK, Grossman RG (1992) Cerebral blood flow and outcome after head injury. J Neurol Neurosurg Psychiatry 55: 594–603

13. Sheinberg M, Kanter JM, Robertson CS, *et al* (1992) Continuous monitoring of jugular venous oxygen saturation in head-injured patients. J Neurosurg 76: 212–217

Correspondence and Reprints: Claudia Robertson, M. D., Department of Neurosurgery. Baylor College of Medicine, One Baylor Plaza, Houston, TX 77030, U.S.A.

Acta Neurochir (1993) [Suppl] 59: 102–106

Monitoring of Jugular Venous Oxygen Saturation in Comatose Patients with Subarachnoid Haemorrhage and Intracerebral Haematomas

A. von Helden, G.-H. Schneider, A. Unterberg, and **W. R. Lanksch**

Department of Neurosurgery, University Hospital Rudolf Virchow, Free University of Berlin, Federal Republic of Germany

Summary

To prevent secondary cerebral ischaemia in comatose patients it would be of great importance to assess cerebral blood flow. Recently monitoring of the jugular venous oxygen saturation ($SJVO_2$) has been shown to continuously evaluate cerebral oxygenation and to estimate cerebral blood flow. While most of these studies have dealt with severely head injured patients, we investigated cerebral oxygenation in 50 comatose patients due to an intracerebral haematoma (n = 14), subarachnoid haemorrhage (n = 12) and severe head injury (n = 24). In these groups of patients, the reaction of $SJVO_2$ to hyperventilation and to lowering of blood pressure was studied. Moderate hyperventilation from 35 to 28 mmHg resulted in a significant decrease of $SJVO_2$ in all groups. A critical $SJVO_2$ between 50 and 55% was found in one half of the patients studied, a pathological $SJVO_2$ below 50% was seen in 23% of the cases. Lowering of arterial blood pressure within the limits of autoregulation resulted in decreases of $SJVO_2$ in patients with intracerebral haematomas only. 55% of these patients showed signs of insufficient cerebral oxygenation. Furthermore the frequency of spontaneous desaturation episodes was studied retrospectively and comparison made between the different groups. These episodes were found more frequently in patients with intracerebral haematomas compared to patients with severe head injury.

In conclusion, monitoring of jugular venous oxygen saturation is a valuable tool for detecting and treating insufficient cerebral oxygenation in comatose patients following intracerebral haemorrhage, subarachnoid haemorrhage and severe head injury.

Keywords: Jugular venous oxygen saturation ($SJVO_2$); cerebral ischaemia; hyperventilation; cerebral perfusion pressure; desaturation episodes.

Introduction

Prevention of secondary cerebral ischaemia is one of the main objectives in the management of comatose patients. Continuous monitoring of intracranial pressure, mean arterial pressure and resulting cerebral perfusion pressure are established as standard methods to secure adequate cerebral perfusion. Assessment of cerebral blood flow at the bedside is technically difficult and yields intermittent values only.

In 1989 C. S. Robertson and coworkers reported a close relationship between the arterio-jugular venous difference of oxygen ($AVDO_2$) and cerebral blood flow[15]. Furthermore, the development of spectroscopic catheters enabled one to continuously measure oxygen saturation in the jugular bulb. Thus, monitoring of cerebral oxygenation became feasible. This method has been used by several groups[1–8, 16, 17, 19, 20].

While the majority of previous studies have dealt with severely head injured patients, the present study investigates cerebral oxygenation in comatose patients due to intracerebral haematomas (ICH) and subarachnoid haemorrhage (SAH grade 4 and 5 according to Hunt and Hess) in comparison to severely head injured patients.

Special emphasis was put on the influence of hyperventilation and moderate lowering of arterial blood pressure on jugular venous oxygen saturation. Furthermore, frequency and causes of spontaneous episodes of desaturation according to Robertson *et al.* were analyzed[16, 17].

Material and Methods

Patient Population

50 comatose patients were studied. 14 patients had suffered ICH, 12 had SAH and 24 patients had had a severe head injury. The mean age in the various groups was between 45 and 60 years, the patients with ICH and SAH being older then the group with severe head injury (Table 1).

Management

All patients were managed by a standard protocol including early intubation and ventilation to ensure an arterial $pO_2 >$

100 mmHg and a pCO_2 of 30–35 mmHg, evacuation of space-occupying haematomas, early aneurysm surgery and repeated CT scans.

Intracranial pressure (ICP), mean arterial blood pressue (MAP), cerebral perfusion pressure (CPP), arterial oxygen saturation (SAO_2), endexpiratory CO_2 and jugular venous oxygen saturation ($SJVO_2$) were continuously monitored. Arterial and jugular venous blood samples were intermittently drawn to confirm jugular venous oxygen saturation in vitro and to determine pO_2, pCO_2, pH and lactate content.

Monitoring of Jugular Venous Oxygen Saturation

To measure oxygen saturation in the jugular bulb, fiberoptic catheters of the Oximetrics 3 System were used. The catheters were percutaneously inserted. Correct placement of the catheter tip in the jugular bulb was confirmed by x-ray. When the study was started, $SJVO_2$ was monitored with a 4 French catheter (Opticath U440, Abbott, Wiesbaden). Later a 5.5 French catheter (Opticath P575EH, Abbott, Wiesbaden) with a small cuff at the tip was used. The latter catheter yielded more stable readings and minimized artifacts. To ensure the accuracy of the fiberoptic measurements, recalibration was routinely performed every 12 hours similarly to others[1, 2, 8].

$SJVO_2$ and Hyperventilation

To study the influence of moderate hyperventilation on cerebral oxygenation, arterial pCO_2 was lowered from a mean of 35 ± 2.6 to 28 ± 2.4 mmHg over a period of 20 minutes by increasing tidal volume and ventilation frequency.

$SJVO_2$ and Lowering of Blood Pressure

The effect of a moderate lowering of blood pressure and thus cerebral perfusion pressure within the limits of cerebral autoregulation on jugular venous oxygen saturation was investigated by infusing urapidil. Mean arterial blood pressure was thus lowered from a mean of 103 ± 2.3 to 84 ± 2.1 mmHg over a period of 20 minutes.

"Desaturation Episodes"

A desaturation episode was defined as a $SJVO_2$ of 50% or less over a period of at least 15 minutes[16, 17]. The frequency of spontaneous desaturation episodes was retrospectively analysed in the different groups of patients.

Results and Discussion

General Aspects

The continuous monitoring of the jugular venous oxygen saturation as a measure of cerebral oxygenation has been studied thoroughly by different groups[1–4, 5–7, 16, 17, 19, 20]. The reliability of the fiberoptic $SJVO_2$ values is good, the correlation with direct measurements by the Co-oximeter after in vivo calibration is between $r = 0.87$ and $r = 0.94$[1, 2, 17]. Our own correlation coefficient of 287 measurements is $r = 0.88$.

Also, the method is safe and complications due to the catheter, in particular thrombosis of the internal jugular vein, have not been seen so far.

Normal/Pathological Values

Before the development of fiberoptic catheters, measurements of the oxygen content in the jugular bulb have been obtained by cannulation and intermittent blood sampling[3, 12–15]. Normal values of the jugular venous oxygen saturation were already established in 1942[12]. $SJVO_2$ values between 55 and 68% were found. Kety and Schmidt reported a normal range of jugular oxygen saturation between 54 and 75%[13].

Garlick and Bihari and Cruz and coworkers were the first to investigate the $SJVO_2$ continuously[3, 5–7, 11]. In patients with severe head injury Cruz graded a decrease of the $SJVO_2$: Grade 1 ranges from 54 to 50%, Grade 2 from 49–45% and Grade 3 below 45% in the presence of normal arterial O_2 content[7].

Like others[17] we defined a $SJVO_2$ between 50 and 55% as critical, below 50% as pathological. If the abnormally low $SJVO_2$ (<50%) lasted for more than 15 minutes, it was defined as a desaturation episode[2, 8, 16, 17]. Moreover, we found an increase of the arterio-jugular venous lactate difference in patients with a $SJVO_2$ of 50% or below indicating insufficient cerebral oxygenation.

Desaturation Episodes

Table 1 presents the frequency of spontaneous desaturation episodes in our patients. During 63 days of monitoring, 59 episodes of desaturation were found by retrospective analysis in the group of patients with ICH. During 65 days of monitoring there were 49 desaturation episodes in patients with SAH. Patients with severe head injury were monitored over a total of 86 days and 53 episodes of insufficient cerebral oxygenation were recorded.

Table 1. Desaturation Episodes in Comatose Patients ($SJVO_2 \leq$ 50% for more than 15 min)

	ICH	SAH	SHI
No of cases	14	12	24
Age	60	50,5	45
Range	(32–82)	(27–68)	(4–79)
Frequency of d.e.	59	49	53
Days of monitoring	63	65	86

ICH	intracerebral haematoma
SAH	subarachnoid haemorrhage
SHI	severe head injury

54% of the severely head injured patients had at least one desaturation episode during monitoring. In 93% of the patients with intracerebral haematoma and in 91% of the patients with SAH one or more desaturation episode was observed.

Episodes of desaturation are frequently reported in patients with severe head injury. Cruz found episodes of abnormally low $SJVO_2$ in almost one half of hyperventilated patients in the presence of normal arterial oxygen, a state he calls cerebral oligaemic hypoxia[7].

Sheinberg *et al.* described 33 confirmed episodes of jugular venous oxygen desaturation in 20 of 45 head injured patients during 1 to 11 days of monitoring[17]. Prakash *et al.* reported desaturation episodes in 48% of their head injured patients during 5 to 7 days of monitoring[16].

Our results are in line with these findings. In addition we observed a considerably higher incidence of desaturation episodes in patients with intracerebral and subarachnoid haemorrhage.

Since there is evidence that the frequency of desaturation episodes is related to neurological outcome, it is important to analyze the reasons for these events[16, 17].

Causes of Cerebral Maloxygenation

A cause of desaturation could be traced in about 65%. Almost 30% of the desaturation episodes were most probably caused by hyperventilation. This corresponds to other studies, where the majority of desaturation episodes was associated with a decrease in pCO_2 or cerebral perfusion pressure[16, 17, 19]. The potentially deleterious influence of hypocapnia on cerebral oxygenation has been suspected for a long time. In 1946 Kety already had strong evidence that the potent method of hyperventilation for treating elevated intracranial pressure might lower CBF to levels critical for sufficient oxygen delivery. Experimental data suggest that intense hyperventilation results in a decrease of PCr as a sign of ischaemia, when CBF is already compromised by intracranial hypertension[18]. Various studies on $SJVO_2$ confirmed the potentially negative effect of hypocapnia on cerebral oxygenation[2-5, 7, 8, 16-19].

The importance of a sufficient cerebral perfusion pressure for an injured brain is well documented by experimental[9] and clinical studies[2, 4, 10, 16, 17, 19].

In this study, patients with ICH were found to be even more susceptible to hypotensive episodes than patients suffering from head injury and SAH. One reason could be an elevated autoregulation threshold in these usually hypertensive patients. They may require a higher CPP to secure sufficient tissue perfusion.

In order to analyze the influence of these two parameters in more detail, the reaction of $SJVO_2$ to hyperventilation and moderate lowering of blood pressure was studied systematically.

Influence of Hyperventilation on $SJVO_2$

Figure 1a summarizes the reaction to moderate hyperventilation in the group of patients with intracerebral haematomas.

All patients but one showed a significant decrease of $SJVO_2$ after the pCO_2 was lowered from a mean of 35 to 28 mmHg. In 5 of the 8 cases studied $SJVO_2$ fell below 55%, in one patient even below 50%.

Comparable results were found in patients after severe head injury (Fig. 1b). The pCO_2 decrease from 36 to 29 mmHg was associated with a $SJVO_2$ below 55% in 6 of 13 cases, in 2 patients jugular

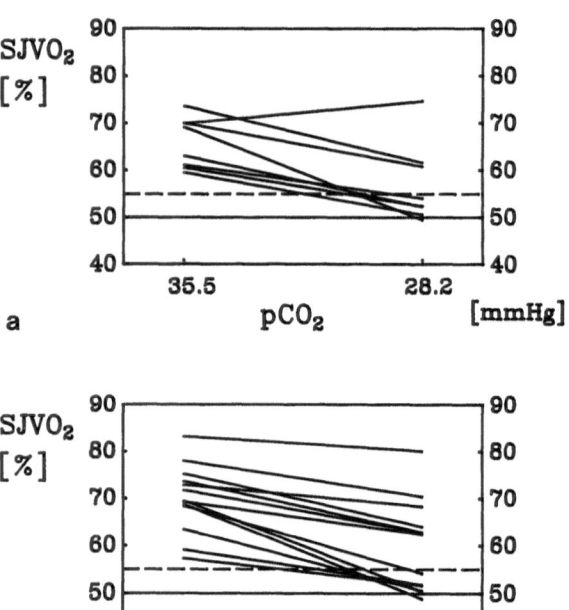

Fig. 1.(a) Reaction of $SJVO_2$ to moderate hyperventilation in patients with ICH (n = 8). All but one patient showed a significant decrease of $SJVO_2$. In 5 cases the saturation fell below 55%, in one patient even below 50%. (b) Reaction of $SJVO_2$ to moderate hyperventilation in patients with severe head injury (n = 13). In 6 cases, $SJVO_2$ fell below 55%, in 2 patients $SJVO_2$ was 50% or below

venous oxygen saturation fell to 50% or below. Patients with SAH reacted to hyperventilation accordingly.

Taken together, a critical $SJVO_2$ below 55% was seen in more than 50% of the patients during moderate hyperventilation with no major differences between the groups, although the pCO_2 was always kept within a generally accepted range (30–28 mmHg). In almost half of these patients (7 of 15) the $SJVO_2$ was abnormally low (<50%) indicating cerebral maloxygenation. These findings indicate that hyperventilation should be done cautiously and preferably during oximetric monitoring. Accordingly, monitoring of $SJVO_2$ is recommended as a guide to titrate the therapy of intracranial hypertension (e.g. hyperventilation vs. mannitol)[3,5-7].

Significance of Blood Pressure Changes in Respect to SJVO₂

Figure 2 summarizes the results of moderate lowering of blood pressure in patients with ICH and severe head injury. 5 of the 9 patients with ICH showed signs of insufficient cerebral oxygenation after lowering the MAP from 103 to 84 mmHg (Fig.

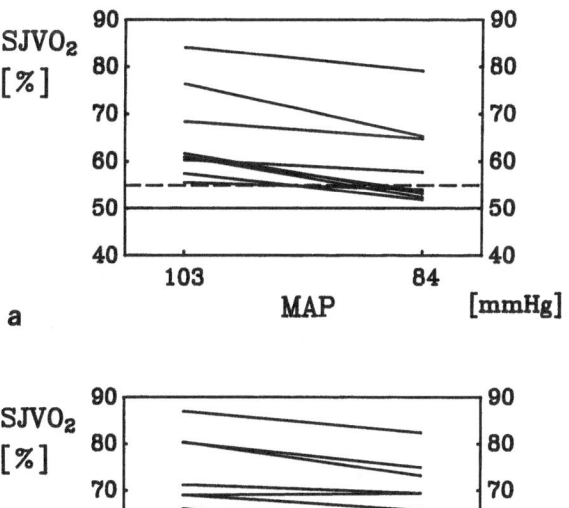

a

b

2a). This is a considerably higher percentage than in the group of head injured patients (Fig. 2b). Following a comparable lowering of blood pressure from 96 to 75 mmHg in head injured patients, $SJVO_2$ fell in some cases only, however never below 50%. These results confirm the observation concerning spontaneous episodes of desaturation, and that patients with ICH react more sensitively to a decrease of cerebral perfusion pressure. As a consequence, a minimum cerebral perfusion pressure of 70 mmHg is recommended to secure sufficient cerebral oxygenation[4,8].

SJVO₂ and Cerebral Vasospasm

Cerebral oyxgenation may also be compromised by cerebral vasospasm. In Fig. 3 (see p. 106), the course of $SJVO_2$ in a patient with severe SAH is shown during development of severe vasospasm.

The patient had suffered SAH grade 4 according to Hunt and Hess from a vertebral artery aneurysm.

On the first by after admission, mean blood flow velocity, measured by transcranial doppler sonography on the middle cerebral artery (MCA), was 88 cm/s. The $SJVO_2$ was around 60% (Fig. 3a). The following day, flow velocity in the MCA increased to 130 cm/s and the $SJVO_2$ decreased simultaneously below 55% (Fig. 3b). On the third day after admission, flow velocity in the MCA went up to 160 cm/s (Fig. 3c). Concomitantly, saturation was constantly below 50% and the arteriovenous lactate difference increased. Thereafter, the CT scan of the brain showed signs of infarction and the patient finally died.

This case demonstrates that severe vasospasm can cause global cerebral maloxygenation. This could be treated effectively in some patients by hypertension and volume expansion[20].

Taken together, continuous monitoring of jugular venous oxygen saturation enables one to detect and to treat insufficient cerebral oxygenation in comatose patients following intracerebral haemorrhage, subarachnoid haemorrhage and severe head injury. It is hoped that this method will help to manage the various problems of these patients more adaequately and to achieve a better outcome.

Fig. 2.(a) Reaction of $SJVO_2$ to moderate lowering of blood pressure in patients with ICH (n = 9). In 5 patients the saturation fell below 55%. (b) Reaction of $SJVO_2$ to moderate lowering of blood pressure in patients with severe head injury (n = 8). In no case did oxygen saturation fall below 55%

Acknowledgement

The technical assistance of Ms R. Duisberg and Ms J. Kopetzki is highly appreciated.

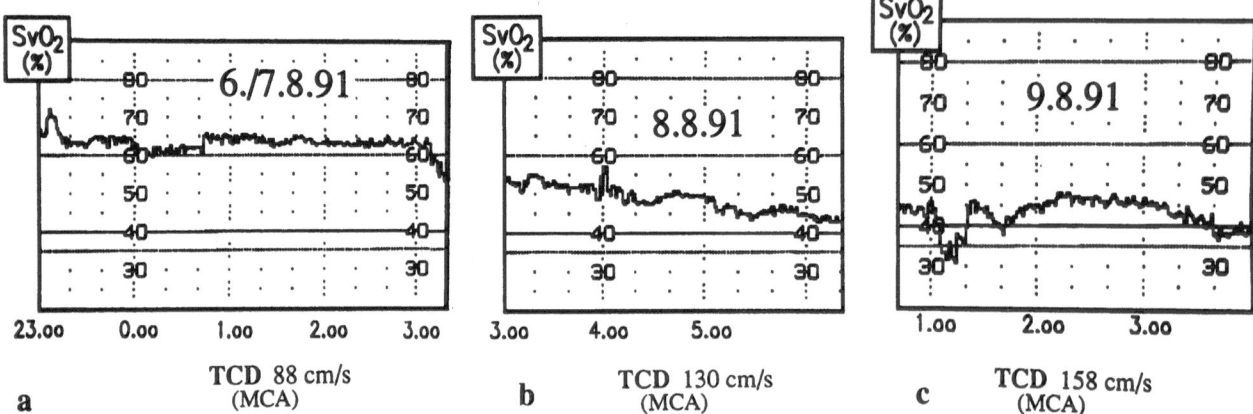

a	TCD 88 cm/s (MCA)	b	TCD 130 cm/s (MCA)	c	TCD 158 cm/s (MCA)

Fig. 3. SJVO$_2$ during development of severe vasospasm in a patient with SAH. (a) First day after admission, SJVO$_2$ around 60%. (b) Second day after admission, increasing flow velocity associated with SJVO$_2$ below 55%. (c) Third day after admission, flow velocity 158 cm/s, SJVO$_2$ permanently below 50%. TCD: Transcranial doppler sonography, MCA: Middle cerebral artery

References

1. Andrews PJD, Dearden NM (1990) Validation of Oximetrics 3 for continuous monitoring of jugular bulb oxygen saturation after severe head injury: Comparison with IL 282 in vitro co-oximeter. Br J Anaesth 64: 393P–394P
2. Andrews PJD, Dearden NM, Miller JD (1991) Jugular bulb cannulation: description of a cannulation technique and validation of a new continuous monitor. Br J Anaesth 67: 553–582
3. Allen SJ, Tonnesen AS, Cruz J, Mackey-Hargardine JR, Miner ME (1984) Cerebral oxygen extraction in patients with head injury. Crit Care Med 4: 230A
4. Chan KH, Miller JD, Dearden NM, Andrews PJD, Midgley S (1992) The effect of changes in cerebral perfusion pressure upon middle cerebral artery blood flow velocity and jugular bulb venous oxygen saturation after severe brain injury. J Neurosurg 77: 55–61
5. Cruz J (1988) Continuous versus serial global cerebral hemometabolic monitoring. Application in acute brain trauma. Acta Neurochir (Wien) [Suppl] 42: 35–39
6. Cruz J, Allen SJ, Miner ME (1985) Hypoxic insults in acute brain injury. Crit Care Med 4: 284
7. Cruz J, Miner ME, Allen SJ, Alves WM, Gennarelli TA (1990) Continuous monitoring of cerebral oxygenation in acute brain injury: injection of mannitol during hyperventilation. J Neurosurg 73: 725–730
8. Dearden NM (1991) Jugular bulb venous oxygen saturation in the management of severe head injury. Curr Opin Anaesth 4: 279–286
9. DeWitt DS, Prough DS, Taylor CL, Whitley JM (1992) Reduced cerebral blood flow, oxygen delivery and electroencephalographic activity after traumatic brain injury and mild hemorrhage in cats. J Neurosurg 76: 812–821
10. Finnerty AF, Within L, Fazehas JF (1954) Cerebral hemodynamics during cerebral ischaemia induced by acute hypotension. J Clin Invest 33: 1227–1232
11. Garlick R, Bihari D (1987) The use of intermittent and continuous recordings of jugular venous bulb oxygen saturation in the unconscious patient. Scand J Clin Lab Invest 47 [Suppl 188]: 47–52
12. Gibbs EL, Lennox WG, Nim LF, *et al* (1942) Arterial and cerebral venous blood. Arterio-venous difference in man. J Biol Chem 144: 325–332
13. Kety SS, Schmidt CF (1948) The nitrous oxide method for the quantitative determination of cerebral blood flow in man, theory, procedere and normal values. J Clin Invest 27: 476–483
14. Robertson CS, Grossman RG, Goodman JC, Narayan RK (1987) The predictive value of cerebral anaerobic metabolism with cerebral infarction after head injury. J Neurosurg 67: 361–368
15. Robertson CS, Narayan RK, Gokaslan ZL, Pahwa R, Grossman RG, Caram P, Allen E (1989) Cerebral arteriovenous oxygen difference as an estimate of cerebral blood flow in comatose patients. J Neurosurg 70: 222–230
16. Prakash S, Robertson CS, Narayan RK, Grossman RG, Hayes C (1992) Transient jugular venous oxygen desaturation and neurological outcome in patients with severe head injury. J Neurosurg 76: 398A
17. Sheinberg M, Kanter MJ, Robertson CS, Contant CF, Narayan RK, Grossman RG (1992) Continuous monitoring of jugular venous oxygen saturation in head-injured patients. J Neurosurg 76: 212–217
18. Sutton LN, McLaughlin AC, Dante S, Kotapka M, Sinwell T, Mills E (1990) Cerebral venous oxygen content as a measure of brain energy metabolism with increased intracranial pressure and hyperventilation. J Neurosurg 73: 927–932
19. Unterberg A, Schneider GH, von Helden A, Lanksch WR (1992) Zerebrovenöse Oximetrie – Messungen in der Vena jugularis interna. In: Olthoff D (ed) Die kontinuierliche Überwachung der gemischtvenösen und organvenösen O$_2$-Sättigung beim kritisch Kranken – Aktueller Stand und Perspektiven. Abbott, Wiesbaden, pp 3.1–3.13
20. Unterberg A, Gethmann J, von Helden A, Schneider GH, Lanksch WR (1992) Treatment of cerebral vasospasm with hypervolemia and hypertension. In: Piscol K, Klinger M, Brock M (eds) Advances in neurosurgery, Vol 20. Springer, Berlin Heidelberg New York Tokyo, pp 198–201

Correspondence and Reprints: Dr. A. von Helden, Department of Neurosurgery, University Hospital Rudolf Virchow, D-13353 Berlin, Federal Republic of Germany.

Acta Neurochir (1993) [Suppl] 59: 107–112

Influence of Body Position on Jugular Venous Oxygen Saturation, Intracranial Pressure and Cerebral Perfusion Pressure

G.-H. Schneider, A. v. Helden, R. Franke, W. R. Lanksch, and **A. Unterberg**

Department of Neurosurgery, Rudolf Virchow Medical Center, Free University of Berlin, Federal Republic of Germany

Summary

Elevation of the head as a common practice to reduce raised intracranial pressure (ICP) has been discussed controversially of late. Some investigators were able to show that besides lowering ICP head elevation may also reduce cerebral perfusion pressure (CPP). For a new evaluation of optimal head position in neurosurgical care it would be of importance to know the influence of body position on cerebral perfusion.

We therefore employed continuous jugular venous oximetry, monitoring cerebral oxygenation, to study the effect of 0°, 15°, 30°, and 45° head elevation on ICP, CPP and jugular venous oxygen saturation (SJVO$_2$) in 25 comatose patients with reduced intracranial compliance.

As expected, head elevation significantly reduced ICP from 19.8 ± 1.3 mmHg at 0° to 10.2 ± 1.2 mmHg at 45°. Already at 30° 92% of the possible effect on ICP was detected. There was no statistically significant change in CPP and SJVO$_2$ associated with varying head position. Individual reactions of CPP to changes in head position, however, were quite unpredictable.

The data suggest that an individual approach to head elevation is to be prefered. A moderate head evelation between 15° and 30° significantly reduces ICP and, in general, does not impair cerebral perfusion. Jugular venous oximetry may be used to optimize ICP, CPP and cerebral oxygenation.

Keywords: Jugular venous oxygen saturation; head elevation; intracranial pressure; cerebral perfusion pressure.

Introduction

Following brain trauma, secondary factors such as raised intracranial pressure (ICP), hypotension, and others, may compromise substrate and oxygen delivery and thus result in secondary brain damage. Traditional patient management in neurosurgical units, therefore, has been directed towards lowering of elevated intracranial pressure and stabilization of cerebral perfusion pressure (CPP) since this was shown to decrease mortality and to improve neurological outcome following severe head injury[13, 22].

In order to control raised ICP the first step in a series of standard procedures is to elevate the patient's head. Though it is generally agreed that this practice leads to a significant reduction of ICP[4, 6, 10, 18], it may also lower CPP, an important prerequisite of sufficient cerebral perfusion[4, 6, 18]. The practice of elevating the head to reduce ICP has thus repeatedly been questioned in recent years[18, 21].

It still remains unclear which body position is the most beneficial for patients with compromised intracranial compliance. Most efforts so far have been directed to continuously monitoring CPP in order to maintain the perfusion pressure above 60 mmHg considered adequate for cerebral perfusion[2]. But even the maintenance of an "adequate" CPP by management of mean arterial blood pressure and ICP is no guarantee for an adequate CBF. Therefore, studies on the value of head elevation in neurosurgical care should try to monitor cerebral blood flow (CBF) itself.

Recently, Feldman *et al.*, using the Kety-Schmidt technique measured CBF during 0° and 30° head elevation in head-injured patients. They found no statistically significant changes either in CPP or CBF[7].

An alternative method to evaluate the quality of CBF is to monitor jugular venous oxygen saturation (SJVO$_2$) as an estimate[5, 16] of cerebral oxygenation. Contrary to the difficult and only intermittent CBF measurements jugular bulb oximetry is a practical bedside method.

To shed more light on the issue of optimal head positioning, we have studied the effect of different degrees of head elevation on ICP, CPP and SJVO$_2$ in comatose patients with intracranial hypertension.

Table 1. *Jugular Venous Oxygen Saturation and Body Position: Patients' Characteristics*

Admitted for:	25 comatose patients severe head injury (n = 17), subarachnoid haemorrhage (n = 5), or intracerebral haemorrhage (n = 3)
GCS on admission:	6 [4–8]
Sex:	m/f = 18/7
Age:	48 ± 3 yrs [20–79]
Baseline ICP:	22 ± 3 mmHg [5–60]; HOB at 15° to 30°

Methods

Characteristics of the group in this study are given in Table 1. 25 comatose patients were included in the study. Patients presented with severe head injury (SHI) in 17 cases, with subarachnoid haemorrhage (SAH) in 5 cases, and with intracerebral haemorrhage (ICH) in 3 cases. Among those with head injury cerebral contusions were the leading pathology in 7 patients, acute subdural haematoma in 6 patients, epidural haematoma in 3 cases, and traumatic subarachnoid haemorrhage in one patient. SAH caused by aneurysm rupture was graded as III in one and as IV in 4 cases on the Hunt/Hess Scale. Mean Glasgow Coma Score on admission was 6, ranging from 4 to 8. There were 18 male and 7 female patients with a mean age of 48 yrs (ranging from 20 to 79 yrs).

Patients were treated according to international standards which included prompt evacuation of intracranial masses, early aneurysm surgery, and aggressive management of intracranial pressure >20 mmHg (HOB at 15° – 30°, moderate hyperventilation with $paCO_2$ = 30 – 35 mmHg, sedation, CSF drainage, mannitol, and finally barbiturates). ICP was monitored via ventriculostomy in 7 patients, or using intraparenchymal (n = 13) or epidural devices (n = 5). Cerebrovenous oxygen saturation was monitored with fiberoptic catheters in combination with the Oximetrix3 System (Abbott Laboratories, North Chicago, IL, U.S.A.). For the first 8 patients we used the two-way, 4F Opticath model U440. Later, we preferred the five-way, 5.5F pulmonary Opticath P575EH.

This catheter offers the advantage of more stable $SJVO_2$ readings with less artifacts, because an inflatable cuff keeps the tip away from the vessel wall. This is especially helpful when the patient's head is moved as in this study.

All catheters were placed in the internal jugular vein using a percutaneous introducer set (Arrow, Reading, PA, U.S.A.). Correct positioning of the catheter tip in the jugular bulb was checked by X-ray. The catheter tip should be placed high in the jugular bulb to minimize the risk of contamination of cerebrovenous blood from extracranial sources. A sterile catheter sheath enabled further small positional changes of the tip in case of light intensity problems. If possible the catheter was placed on the side of the dominant lesion. The catheter was maintained according to standard care and kept patent with a continuous heparin flush. After preinsertion calibration the measured $SJVO_2$ values were crosschecked on a CO-oximeter (Instrumentation Laboratory, Lexington, MA, U.S.A.) by jugular blood samples drawn from the catheter at regular intervals.

Mean arterial pressure (MAP) was measured in the radial artery and care was taken to keep the transducer at the level of the external auditory canal. CPP was calculated as the difference of MAP and ICP.

Study measurements were performed within 72 hrs after admission of the patients. At the beginning of the protocol the head was elevated to 45°. The head was then lowered by three 15° steps (each lasting 20 min) to the flat body position. Head elevation was measured by flexion at the hips and due care was taken to keep head, neck and spine in a straight line.

At each position ICP, MAP, and $SJVO_2$ were recorded. Arterial and jugular venous blood samples were drawn for determination of blood gases, pH, oxygen saturation as well as lactate concentrations.

Statistical evaluation of data was performed using the Friedman test for paired values.

Results

Figure 1 and Table 2 summarize the essential findings of this study. Compared to the supine position (ICP: 19.8 ± 1.3 mmHg; MAP: 79.9 ±

Table 2. *Effect of Head Elevation on Cerebral and Systemic Parameters*

	Head Elevation [°]			
	45°	30°	15°	0°
ICP [mmHg]	10.2 ± 1.2*	11.0 ± 1.2*	14.3 ± 1.4*	19.8 ± 1.3
MAP [mmHg]	68.9 ± 1.9*	72.3 ± 1.9*	75.7 ± 2.0**	79.9 ± 2.1
CPP [mmHg]	59.9 ± 2.3	61.5 ± 2.0	61.8 ± 2.0	60.3 ± 2.3
$SJVO_2$ [%]	65.1 ± 2.1	66.3 ± 1.9	64.7 ± 1.9	67.2 ± 2.0
p_aCO_2 [mmHg]	34.2 ± 0.7	33.9 ± 0.8	34.0 ± 0.7	33.9 ± 0.7
AVDL [mg/dl]	−0.4 ± 0.2	−0.4 ± 0.2	−0.3 ± 0.2	−0.4 ± 0.2

* $p < 0.001$.
** $p < 0.01$.

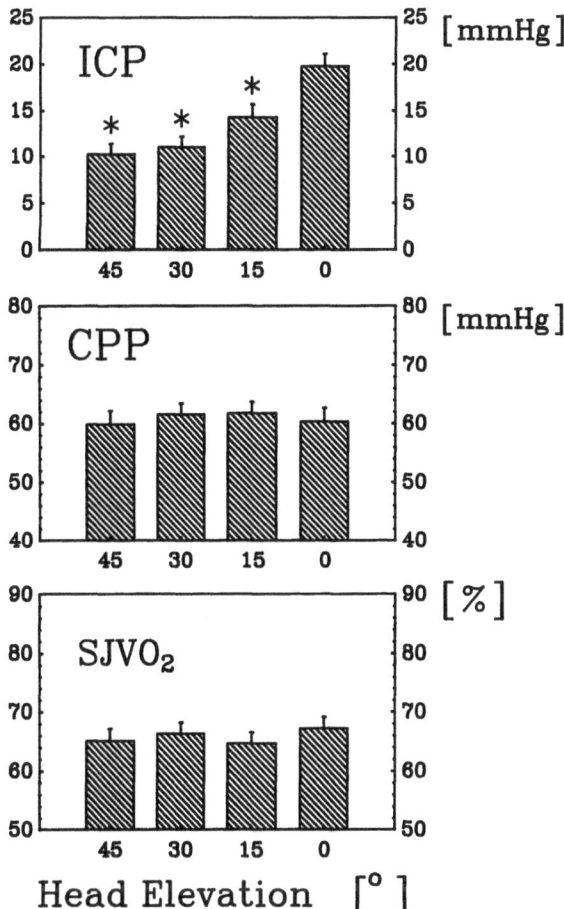

Fig. 1. Effect of body position on intracranial pressure (*ICP*), cerebral perfusion pressure (*CPP*) and jugular venous oxygen saturation (*SJVO$_2$*). While ICP was significantly lowered by head elevation, CPP and SJVO$_2$, as an estimate for CBF, remained unchanged

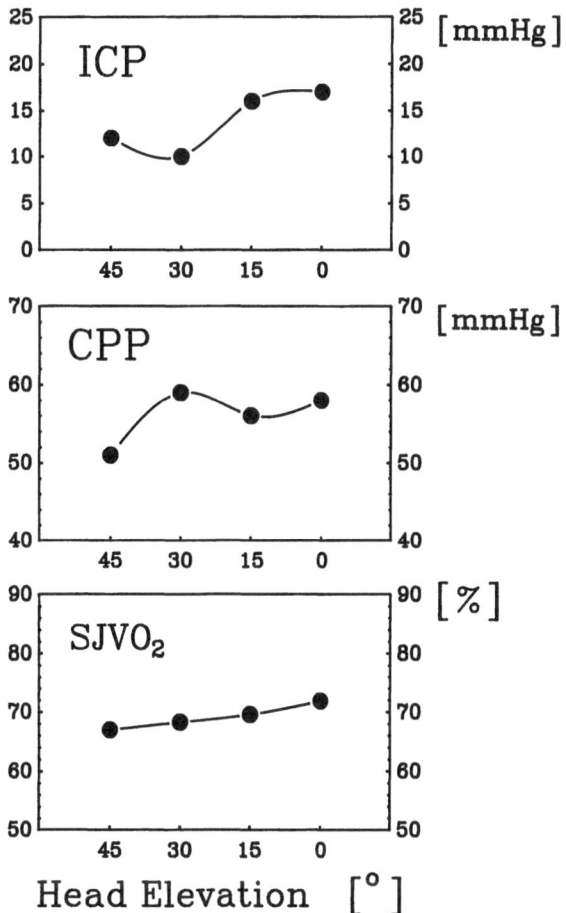

Fig. 2. Response of *ICP*, *CPP* and *SJVO$_2$* to head elevation in Patient A who had suffered a severe head injury. *ICP* was lowest in the 30° position. At 45° head elevation *CPP* dropped to a critical level. *SJVO$_2$* gradually increased when the patient was lowered to the horizontal position. For this patient 30° appears to be the optimal position

2.1 mmHg) the mean values of ICP and MAP (measured at the level of the external auditory meatus) were significantly lower at 45° (ICP: 10.2 ± 1.2 mmHg; MAP: 69.8 ± 1.9 mmHg), 30° (ICP: 11.0 ± 1.2 mmHg; MAP: 72.3 ± 1.9 mmHg) and 15° (ICP: 14.3 ± 1.4 mmHg; MAP: 75.7 ± 2.1 mmHg) of head elevation. Thus, almost 60% of the total reduction of ICP was achieved by a moderate head elevation of 15°. 92% of the possible effect on ICP was evident at 30° head elevation.

Individual data showed that the extent of ICP reduction effected by head elevation depended on the initial ICP level. A positive correlation between the ICP level in the horizontal position and the amount of change at 15° and 30° head elevation was found.

As the reduction of ICP and MAP grossly parallelled each other the mean values for CPP were

nearly unaffected during the manoeuvre: CPP was 60.3 ± 2.3 mmHg at 0° and 59.9 ± 2.3 mmHg at 45°.

Also, the mean value of jugular venous oxygen saturation remained unaffected by head elevation (65.1 ± 2.1% at 45°, 66.3 ± 1.9% at 30°, 64.7 ± 1.9% at 15° and 67.2 ± 2.0% lying flat); nor did other parameters like the paCO$_2$ or the arteriovenous difference of lactate (AVDL) show any statistically significant changes during positional changes.

Although CPP was unaffected by varying head positions when the data for all 25 patients were averaged, the individual response of CPP to head elevation was unpredictable: Comparing the individual CPP values at 0°, e.g., CPP was higher in 43%, the same in 10%, and lower in 47%.

Figures 2 and 4 illustrate two different responses (Case A and B, respectively):

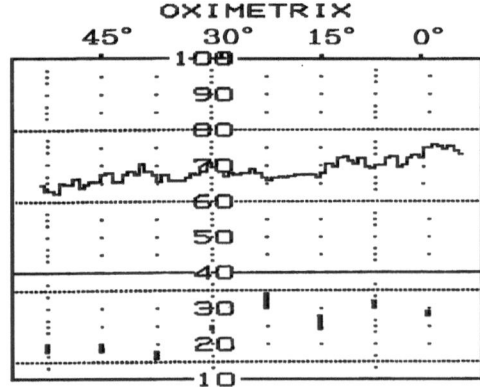

Fig. 3. Original Oximetrix[3] tracing of the positional study of Patient A

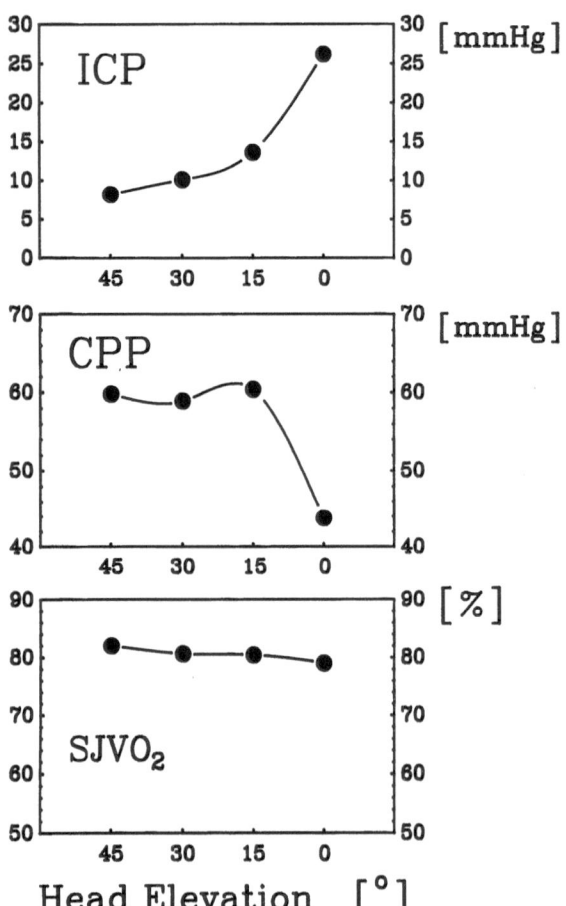

Head Elevation [°]

Fig. 4. Effect of head elevation on *ICP, CPP* and *SJVO₂* in Patient B following SAH. *ICP* progressively increased with lowering of the head. *CPP* suffered a steep drop when the patient reached the horizontal position. Despite the very low *CPP*, jugular venous oxygen saturation did not indicate insufficient cerebral oxygenation. The manoeuvre was nevertheless immediately stopped because of the high *ICP* and the low *CPP*

Patient A (Fig. 2), a 39 year old male, had suffered a severe head injury with multiple contusions and a right subdural haematoma which was evacuated. In this patient, ICP was lowest in the 30° position and started to rise as soon as the head was elevated to 45°. This rise in ICP was obviously caused by the acute drop of CPP to critical levels (6, 7) when the body was positioned at 45°. When the patient was gradually lowered to the horizontal position $SJVO_2$ slowly increased from 67% to 72% paralleling the rise of CPP. For this patient a head elevation of 30° appears to be the optimal position becaue of the lowest ICP and the best CPP. Figure 3 is an original tracing of the positional study of this patient.

Patient B (Fig. 4), a 44 year old female, was admitted due to a SAH from a basilar aneurysm. When the head was gradually lowered, a progressive increase in ICP from 8 to 26 mmHg became obvious. While CPP was the same between 45° and 15°, it suffered a steep drop to 44 mmHg when the patient reached the horizontal position. Despite the very low CPP cerebrovenous oxygen saturation stayed well above 70%, but the manoeuvre was immediately stopped because of the low CPP and high ICP.

Further analysis of data revealed a subgroup of 7 patients (28%) with a characteristic ICP pattern: ICP progressively declined from the horizontal position to 30° elevation to rise again significantly when the head was elevated above 30°. This subgroup could be distinguished by a significant drop of CPP above 30° head elevation. Cerebrovenous oxygen saturation, however, remained unaffected in this group. In contrast to these 7 patients all others showed a steady decline of ICP up to 45° head elevation without a change in CPP.

Discussion

Our results reaffirm other studies that elevation of the head leads to a quick and significant reduction of intracranial pressure[4, 6, 7, 10, 18]. There appear to be two major mechanisms for this ICP reduction. One is the facilitation of venous outflow[11, 14, 15]. Central venous pressure and jugular venous pressure markedly decline with head elevation[8] which in turn leads to a reduction in cerebral blood volume (CBV).

The possibility of hydrostatic cerebrospinal fluid (CSF) displacement into the spinal subarachnoid space has been suggested as another mechanism for ICP reduction[10, 12].

The effectiveness of these mechanisms to reduce increased ICP obviously depends upon the state of intracranial compliance and the relative amount of fluid (either blood or CSF) that can be displaced from the intracranial compartment. As mirrored by our data, the higher the patient's baseline ICP, the more effective is the reduction in ICP by head elevation.

While all our patients showed a decline in ICP between 0° and 30°, other investigators report varying responses of ICP to head elevation[7, 18], though.

When studying ICP and CPP as a function of head position in patients with raised ICP, Rosner et al. found CPP to be reduced by any degree of head elevation[18]. They argued that a fall in CPP results in vasodilation which will increase CBV and ICP. If ICP increases without an increase in MAP, a self-sustaining cycle is started because the raised ICP will further decrease CPP[19, 20]. Since Rosner et al. noted intracranial pressure waves to be more common in patients with their head elevated[17] and CPP was maximal in the horizontal body position in their study they advocated flat nursing of patients with moderately increased ICP. Despite the obvious importance of CPP, sole management of CPP cannot fully guarantee sufficient cerebral perfusion because of the unknown vascular resistance. Since the primary aim of ICP management is an adequate CBF it seems more apt to monitor CBF to evaluate the practice of head elevation in neurosurgical care. Methods to measure CBF directly are difficult and time consuming. Jugular venous oximetry appears to be a practical bedside method helpful in order to clarify the question of an ideal head position for neurosurgical patients. According to Fick's principle SJVO₂ is proportional to CBF if arterial saturation, cerebral metabolic rate of oxygen, haemoglobin concentration and haemoglobin dissociation curve are constant. A cerebrovenous oxygen saturation below 50% is considered pathological[1, 3].

Inspite of individually different reactions of CPP to head elevation jugular venous oxygen saturation did not significantly change and always stayed within normal limits indicating sufficient global cerebral oxygenation in our patients. This is in accordance with Feldman[7] who found no significant effect of 0° and 30° head elevation on CBF.

Although no significant effect of varying head positions on SJVO₂ was found one must bear in mind that cerebrovenous oximetry is a method for global measurement of cerebral oxygenation. It is therefore conceivable that in patients with higher ICP and less compliance, regional areas of ischaemia may exist, especially in penumbra zones[9]. These areas of local ischaemia may remain undetected by global jugular venous oximetry.

The results of the present study suggest that a moderate head elevation between 15° and 30° still appears to be a practical method to control raised ICP and, in general, does not jeopardize cerebral oxygenation. For purposes of lowering ICP a head elevation above 30° is not necessary.

Since the individual reactions to head elevation were markedly different, an individual approach to head elevation is to be preferred. If possible, continuous monitoring of jugular venous oxygen saturation (or other methods, like near-infrared spectroscopy) should be employed to optimize ICP, CPP and cerebral oxygenation in patients with raised intracranial pressure.

Acknowledgement

The excellent technical assistance of R. Duisberg and J. Kopetzki is highly appreciated.

References

1. Andrews PJD, Piper IR, Dearden NM, et al (1990) Secondary insults during intrahospital transport of head-injured patients. Lancet 335: 327–330
2. Chan KH, Miller JD, Dearden NM, et al (1992) The effect of changes in cerebral perfusion pressure upon middle cerebral artery blood flow velocity and jugular bulb venous oxygen saturation after severe brain injury. J Neurosurg 77: 55–61
3. Cruz J, Miner ME, Allen SJ, et al (1990) Continuous monitoring of cerebral oxygenation in acute brain injury: injection of mannitol during hyperventilation. J Neurosurg 73: 725–730
4. Davenport A, Will EJ, Davidson AM (1990) Effect of posture on intracranial pressure and cerebral perfusion pressure in patients with fulminant hepatic and renal failure after acetaminophen self-poisoning. Crit Care Med 18: 286–289
5. Dearden NM (1991) Jugular bulb venous oxygen saturation in the management of severe head injury. Curr Opin Anaesthesiol 4: 279–286
6. Durward QJ, Amacher AL, Del Maestro RF, et al (1983) Cerebral and cardiovascular responses to changes in head elevation in patients with intracranial hypertension. J Neurosurg 59: 938–944
7. Feldman Z, Kanter MJ, Robertson CS, et al (1992) Effect of head elevation on intracranial pressure, cerebral perfusion pressure, and cerebral blood flow in head-injured patients. J Neurosurg 76: 207–211
8. Gauer OH, Thron HL (1965) Postural changes in the circulation. In: Hamilton WF (ed) Handbook of physiology, Vol 3. Williams and Wilkins, Baltimore, pp 2409–2439
9. Jafar JJ, Crowell RM (1987) Focal ischemic thresholds. In: Wood JH (ed) Cerebral blood flow – physiological and clinical aspects. McGraw Hill, New York, pp 449–457
10. Kenning JA, Toutant SM, Saunders RL (1981) Upright patient positioning in the management of intracranial hypertension. Surg Neurol 15:148–152
11. Magnaes B (1976) Body position and cerebrospinal fluid pressure. Part 1: clinical studies on the effect of rapid postural changes. J Neursurg 44: 687–697
12. Magnaes B (1978) Movement of cerebrospinal fluid within the craniospinal space when sitting up and lying down. Surg Neurol 10: 45–49

13. Marmarou A, Eisenberg HM, Foulkes MA, Marshall LF, Jane A, *et al* (1991) Impact of ICP instability and hypotension on outcome in patients with severe head trauma. J Neurosurg 75: S59–S66
14. Miller JD, Bell BA (1987) Cerebral blood flow variations with perfusion pressure and metabolism. In: Wood JH (ed) Cerebral blood flow – physiological and clinical aspects. McGraw Hill, New York, pp 119–130
15. Potts DG, Deonarine V (1973) Effect of positional changes and jugular vein compression on the pressure gradient across the arachnoid villi granulations of the dog. J Neurosurg 38: 722–728
16. Robertson CS, Narayan RK, Goskaslan ZL, *et al* (1989) Cerebral arteriovenous oxygen difference as an estimate of cerebral blood flow in comatose patients. J Neurosurg 70: 222–230
17. Rosner MJ, Becker DP (1984) Origin and evolution of plateau waves – experimental observations and a theoretical model. J Neurosurg 60: 312–324
18. Rosner MJ, Coley IB (1986) Cerebral perfusion pressure, intracranial pressure and head elevation. J Neurosurg 65: 636–641
19. Rosner MJ (1986) The vasodilatory cascade and intracranial pressure. In: Miller JD, *et al* (eds) Intracranial pressure, Vol 6. Springer, Berlin Heidelberg New York Tokyo, pp 137–141
20. Rosner MJ (1987) Cerebral perfusion pressure: link between intracranial pressure and systemic circulation. In: Wood JH (ed) Cerebral blood flow – physiological and clinical aspects. McGraw Hill, New York, pp 425–448
21. Rosner MJ, Daughton S (1990) Cerebral perfusion pressure management in head injury. J Trauma 30: 933–941
22. Saul TG, Ducker TB (1982) Effect of intracranial pressure monitoring and aggressive treatment on mortality in severe head injury. J Neurosurg 56: 498–503

Correspondence and Reprints: Dr. G.-H. Schneider, Department of Neurosurgery, Rudolf Virchow Medical Center, Free University of Berlin, D-13353 Berlin, Federal Republic of Germany.

Acta Neurochir (1993) [Suppl] 59: 113–118

Continuous Monitoring of Jugular Bulb Oxygen Saturation and the Effect of Drugs Acting on Cerebral Metabolism

R. Bullock, L. Stewart, C. Rafferty, and **G. M. Teasdale**

University Department of Neurosurgery, Institute of Neurological Sciences, Glasgow, Scotland, U.K.

Summary

The laser absorption spectrophotometric technique was used to continuously monitor jugular bulb oxygen saturation (SjO_2), and thus to calculate arteriovenous oxygen differences ($AVDO_2$), in three subgroups of intensively monitored, severely head injured patients. We have used this data to address two questions: 1. How do cerebral and systemic haemodynamic changes affect SjO_2 and $AVDO_2$; and 2. Can $ADVO_2$ measurements be used to detect *therapeutic* changes in brain metabolism, in response to drugs. The major haemodynamic factor affecting SjO_2 and $AVDO_2$ was intracranial pressure (ICP). Increases in ICP were associated with concomitant increases in SjO_2, and decreases in $AVDO_2$, suggesting cerebral hyperaemia in response to ICP waves. Systemic changes were less frequent, but potent influences on SjO_2. The short acting anaesthetic agent propofol produced a marked increase in SjO_2 (decrease in $AVDO_2$ to below the normal range) which became less marked with time. A new high affinity glutamate antagonist produced no change in SjO_2.

With rigorous attention to technical factors, and exclusion of extra and intracranial haemodynamic effects, SjO_2 monitoring may be a useful "surrogate end point" for the effect of drugs acting on brain metabolism.

Keywords: Severe head injury; brain metabolism; arteriovenous oxygen difference; neuroprotective drugs.

Introduction

Intermittent measurement of cerebral arteriovenous oxygen difference ($AVDO_2$) has been used by a variety of authors to improve our understanding of the pathophysiological processes which occur acutely after severe head injury[2,3,4,6]. $AVDO_2$ measurements, particularly when combined with cerebral blood flow (CBF) and intracranial pressure (ICP) measurements, can allow the diagnosis of cerebral hyperaemia (luxury perfusion) which occurs at some time during the clinical course of up to 85% of young head injured patients[3]. When $AVDO_2$ measurements are used in conjunction with assessments of the cerebral arteriovenous lactate content (AVDL) critical cerebral ischaemia can be determined thus providing an understanding of the factors which may be responsible for secondary insults during head injury care[6]. Despite these important contributions, however, there has been little evidence to show that $AVDO_2$ measurements have been of value in guiding different forms of *therapy* for head injured patients. This may be because the measurements were intermittent only.

It has recently become possible to *continuously monitor* jugular venous oxygen saturation (SjO_2) using laser absorption spectrophotometry techniques (Oximetrix 3 System, Abbott Laboratories, Kent, U.K.) which allows $AVDO_2$ to be calculated if SaO_2 is unchanged. The possibility now exists that these continuous techniques may be useful to allow modification of therapy to optimise the cerebral metabolic milieu in the light of continuously available oxygen saturation data.

Robertson *et al.*, Cruz *et al.*, and our own group (Rafferty *et al.*) have shown that the validity of the Oximetrix measurements is crucially dependent upon technical factors such as the positioning of the catheter, attention to light intensity levels and calibration of the system[2,5,6]. In various different studies, between 15 and 50% of SjO_2 Oximetrix readings have been found to be unreliable due to these factors. At the Institute of Neurological Sciences in Glasgow, we have employed SjO_2 monitoring in over 40 severely head-injured patients with a monitoring period ranging between two and seven days (mean 3.6 days). We have used the data thus acquired to address two questions: 1. How do acute cerebral haemodynamic events affect SjO_2; and 2.

Can SjO$_2$ monitoring be used to detect changes in cerebral metabolism induced by drug therapy?

Patients and Methods

Patients in coma with severe head injury in Glasgow are monitored in accordance with an established protocol as follows. Whenever an intradural haematoma is present, and where the CT scan demonstrates features of raised intracranial pressure, such as shift of the midline structures (<5 mm), effacement of basal cisterns or multiple cerebral contusions with oedema, intracranial pressure monitoring is performed (currently using the Camino 420 intraparenchymal fibreoptic system by Camino Laboratories, San Diego, California, U.S.A.). In those patients with raised intracranial pressure, artificial ventilation is continued initially for 24 to 48 hours or longer, depending on intracranial pressure trends and responses to therapy. If ICP remains >25 mmHg, ventilation is continued.

Continuous AVDO$_2$ monitoring is instituted for those patients in whom intracranial pressure remains raised, despite ventilation and for those who are hypoxic, and require ventilation for pulmonary reasons, with a coexistent significant head injury. SjO$_2$ monitoring is performed using the Oximetrix fibreoptic spectrophotometric system. The fibreoptic catheter is inserted through a 14 gauge sheath inserted into the internal jugular vein. After excluding a coagulopathy, Seldinger techniques are used to position the tip of the Oximetrix catheter at or slightly above the jugular bulb using the right side for preference, if jugular compression causes no change in ICP. Its position is checked using a lateral skull x-ray. Calibration of the Oximetrix system is performed in vitro prior to insertion of the catheter, and in vivo, at least eight hourly. Calibration checks are performed by measuring the oxygen content of jugular bulb venous blood using a co-oximeter. The oxygen content of arterial blood drawn concomitantly from an indwelling radial artery catheter is used to calculate AVDO$_2$ values[5,6].

The Oximetrix SjO$_2$ data is continuously plotted onto chart paper, together with ICP and CPP traces. The data is also captured on a multichannel computer system (Wolter Graphtech, Essen, Federal Republic of Germany) and stored on optical discs. In addition, mean arterial blood pressure, derived cerebral perfusion pressure, arterial oxygen saturation (SaO$_2$), and a 12 channel EEG spectrum are continuously logged for data analysis.

In order to assess haemodynamic changes, all paper traces were visually checked, and any changes in SjO$_2$ >5% were noted, together with all ICP changes >10 mmHg.

Pharmacological Therapy

Background sedation for patients with severe head injury, who are monitored using the above methods, is by the use of benzodiazepines (Midazolam 0.05–0.2 mg/kg/h) together with Vecuronium (0.05–0.1 mg/kg/h). When analgesia is necessary, fentanyl (0.05–2.0 u gm/kg/h) or morphine (0.5–0.1 mg/kg/h) are used for ventilated patients.

Patients

In a group of 12 patients, propofol (Diprovan, ICI Pharmaceuticals, Macclesfield, U.K.) was administered in doses ranging from 150 to 400 mg/hour, (mean 237 mg/h) as a continuous infusion, in order to control intracranial pressure.

In a further selected subgroup of 15 patients, a new high affinity glutamate antagonist was administered in two bolus doses 24 hours apart.

Results

The Effect of Acute Haemodynamic Events upon SjO$_2$

Extracranial Factors

The largest magnitude of change in SjO$_2$ was seen in response to extracranial events such as changes in oxygen saturation and mean arterial blood pressure. Preoxygenation by manual ventilation (100% oxygen) produced an increase in SjO$_2$ up to 12%. This response was fairly constant irrespective of the baseline SjO$_2$ or clinical status of the patient. Similarly, in one patient an episode of accidental disconnection from the ventilator, with a decrease in SaO$_2$ to 58% produced a massive fall in SjO$_2$ (50% decrease) which was brief and short lasting. This was accompanied by a 25 mmHg increase in ICP.

Intracranial Haemodynamics and SjO$_2$

The clinical features of the patients in whom SjO$_2$ was continuously monitored, with and without administration of putatively metabolically active drugs (the "control" and propofol groups) are shown in Table 1. The number of acute transients in SjO$_2$ per patient and per day ranged from 6 to 35. Transients were more frequent during the first day of monitoring than subsequently during the clinical course. They were more frequent in patients who died and in those who had high intracranial pressure than in those with a favourable outcome (Fig. 2). The most important accompaniments of SjO$_2$ transients were changes in intracranial pressure (Fig. 1). An increase in SjO$_2$ was far more frequent than a decrease in

Table 1. *Clinical Features in Patients with Continuous SjO$_2$ Monitoring*

	Control (n = 21)	Propofol Therapy (n = 21)
Mean age (range years)	33 (12–66)	23 (3–58)
Initial Glasgow Coma Score (range)	7 (4–10)	5 (3–10)
Haematoma removed	11	4
Mean ICP (first 24 hours) (mmHg)	23 ± 16	20 ± 14
Outcome		
dead		
vegetative	8	6
severe disability		
good outcome		
minimal disability	13	6

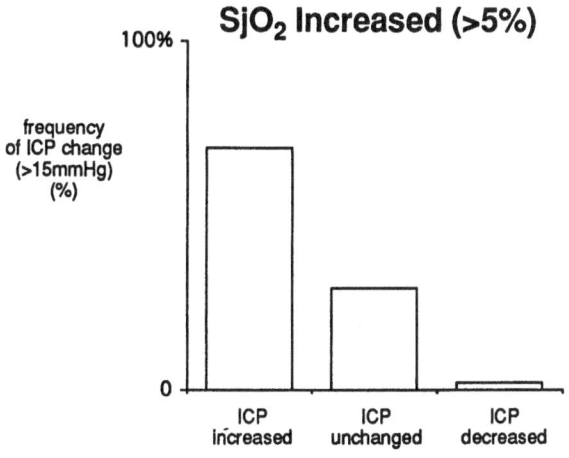

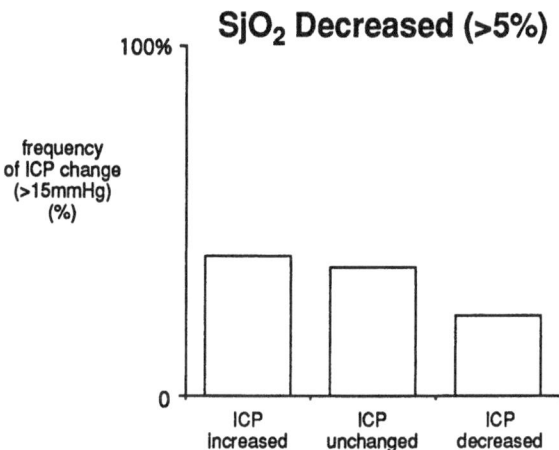

Fig. 1. Changes in SjO_2 in response to ICP

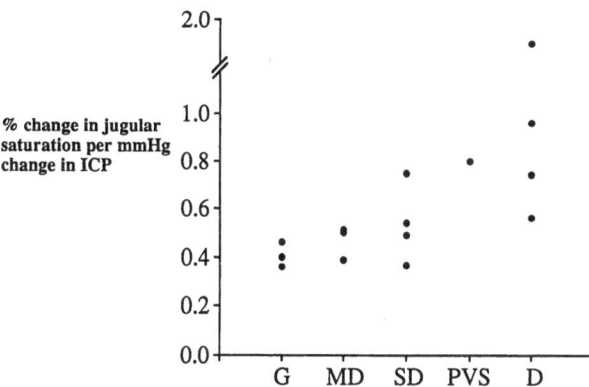

Fig. 2. Relationship between the magnitude of increase in jugular venous oxygen saturation per mm Hg, ICP rise, and outcome at three months after severe head injury (for outcome categories see Table 1)

SjO_2 – 77% increases versus 23% decreases. It may be seen from Fig. 1 that the most frequent accompaniment of an increase in SjO_2 was an increase in intracranial pressure, and this was most usually a consequence of an increase in mean arterial blood pressure. This phenomenon was most frequently seen in patients with a poor outcome, suggesting impaired cerebrovascular autoregulatory capacity such that changes in blood pressure were transmitted directly to the intracranial contents and reflected as an intracranial pressure increase (Fig. 2).

In just over a quarter of the patients, ICP did not change when SjO_2 increased, and in only 2% of the SjO_2 transients was a concomitant *decrease* in intracranial pressure noted. Decreases in SjO_2 were less frequent, and most usually associated with an increase in intracranial pressure. However, the intracranial pressure responses associated with an SjO_2 decrease were much more heterogeneous (Fig. 1).

Effect of Propofol Therapy upon $AVDO_2$

The clinical features of the group of 12 patients who were treated with propofol are shown in Table 1 – there were no significant differences in age, coma score, ICP or outcome in this group when compared with the similar subset of patients who underwent $AVDO_2$ monitoring but who were managed without propofol. The effect of propofol upon $AVDO_2$ is shown in Fig. 3. There was a marked decrease in $AVDO_2$ after commencement of propofol (mean 6.5 ± 5 ml/dl down to 3 ± 1 ml/dl) over the first four hours of propofol administration, but this change was not significant because of the heterogeneous pre-propofol $AVDO_2$ values. During the subsequent propofol therapy period, $AVDO_2$ values progressively increased, although they remained below the pre-propofol values throughout the two day period for which data is available.

The effect of propofol infusion upon cerebral perfusion pressure may be seen in Fig. 4. In three patients, inotropic agents were added, after propofol had been commenced, and they are deleted from this analysis. When propofol was used alone, the effect upon cerebral perfusion pressure was slightly beneficial, particularly when compared with the cerebral perfusion pressure changes in the subgroup who did not receive propofol or inotropic agents (n = 6).

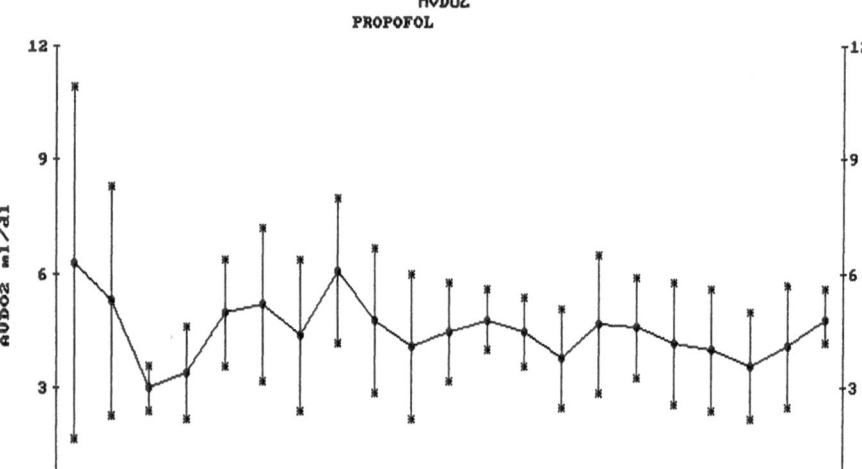

Fig. 3. AVDO$_2$ changes measured over 40 hours after commencement of propofol infusion in 12 patients

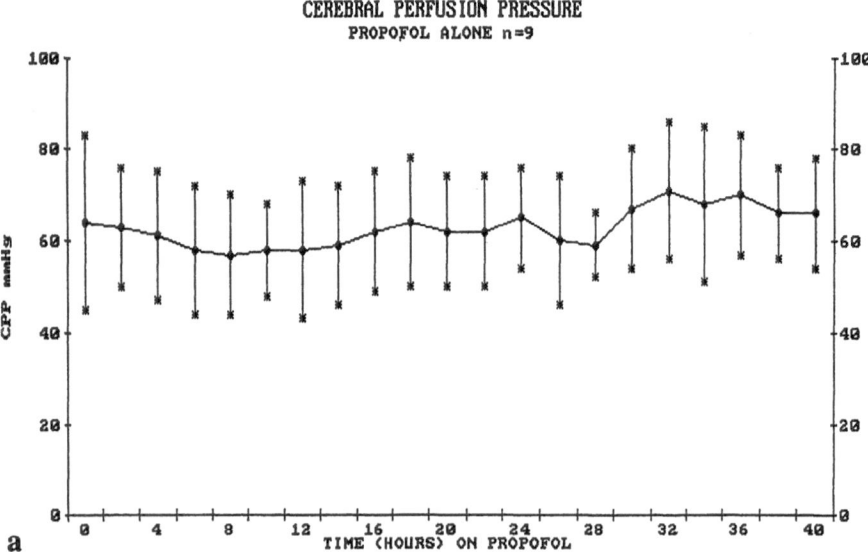

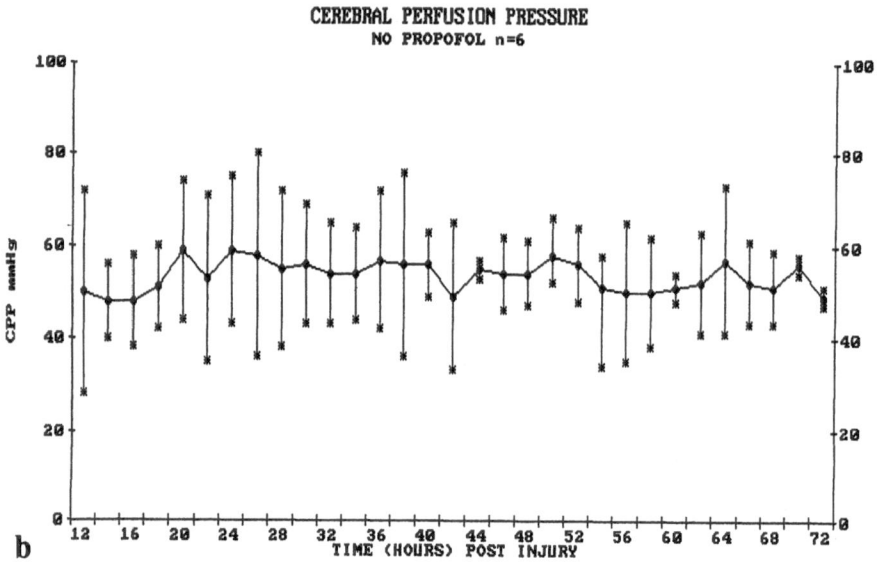

Fig. 4. Cerebral perfusion pressure in patients given propofol (a) and in the "control group" to whom propofol was not administered (b)

Effect of a Glutamate Antagonist upon $AVDO_2$ and Haemodynamic Parameters

All the patients who received a glutamate antagonist underwent $AVDO_2$ and ICP – CPP monitoring for two hours prior to, and for 72 hours after treatment. No significant changes were seen in $AVDO_2$, MABP, ICP, CPP, or transcranial Doppler velocities in the middle cerebral artery, after treatment with this agent.

Discussion

This study clearly demonstrates that $AVDO_2$ is a highly dynamic parameter, which fluctuates in response to changes in intracranial haemodynamic events. Although $AVDO_2$ transients were also seen to occur in response to extracranial factors such as rapid changes in blood pressure or arterial oxygen saturation, by far the commonest event associated with rapid $AVDO_2$ changes were changes in intracranial pressure. The commonest of these changes consisted of an *increase* in intracranial pressure in association with an *increase* in SjO_2 (Fig. 1). In previous literature, prior to the availability of continuous SjO_2 monitoring, the emphasis has been upon detection of cerebral ischaemia using the $AVDO_2$. Our data has indicated that decreases in SjO_2 are much less frequent than increases (23% versus 77%) and that in *under half* of these *decreases* in SjO_2, is there an associated increase in intracranial pressure. This implies that an SjO_2 decrease may be more often related to a non-ICP event, such as arterial desaturation.

We postulate that these SjO_2 increases are due to increases in mean arterial blood pressure being transmitted to the intracranial compartment as intracranial pressure waves, in patients in whom autoregulation has been severely impaired, and in whom brain stiffness is increased. In Fig. 2, we have shown that patients who had a favourable outcome generally demonstrated small changes in SjO_2 in response to ICP fluctuations, while those who died manifested the largest fluctuations, suggesting that impaired autoregulation is a factor in this relationship.

Although cerebral blood flow was not measured in this series of patients, we would hypothesise that *cerebral hyperaemia* could explain the relationship between increases in SjO_2 and increases in intracranial pressure, particularly in the face of disturbed cerebrovascular autoregulation. This would accord

with the findings of Obrist and Muizelaar *et al.*[3, 4]. Our data did not, however, suggest that such labile cerebrovascular events were more common in younger patients.

These data indicate that single intermittent measurements of $AVDO_2$ are of limited value, and should not be used for management decisions or classification of patients. In order to obtain a true reflection of the factors affecting cerebral oxygenation, $AVDO_2$ values must be frequently measured and averaged, or alternatively, baseline recordings taken from continuous trends may be used.

In the second part of this study we have assessed the effect of two "metabolically active" agents upon $AVDO_2$. The short acting anaesthetic agent propofol has been shown by other authors to reduce intracranial pressure, and decrease the SjO_2 in a manner which is similar to barbiturates (Thiopentone), although propofol was also hypotensive[1]. Propofol, however, has the advantage that its action is of much shorter duration, and its effects upon cerebral perfusion pressure may be less marked than those of thiopentone. Commencement of propofol infusion produced a marked rise in SjO_2 (a fall in $AVDO_2$) without a change in MCA flow velocity, consistent with a decrease in cerebral metabolic rate. This effect became less marked with time over the two day period for which data were collected, but after propofol administration, $AVDO_2$ remained well below the range for normal values, reported by Obrist *et al.*[4]. These changes were associated with only slight changes in cerebral perfusion pressure, although initially the hypotensive effect of propofol caused CPP to decrease slightly.

The high affinity, post-synaptic glutamate antagonist used as the second "metabolically active" drug in this study produced no effect on $AVDO_2$, ICP, MABP, CPP or middle cerebral artery flow velocity measured by transcranial Doppler. However, because this was a dose escalation study, it is possible that the dose of drug administered was insufficient to influence brain metabolism significantly. It is well known that glutamate antagonists differentially affect different parts of the mammalian brain, with increased metabolism in the limbic system and decreased metabolism in the remaining areas. It may be that the effect of these modulatory changes upon differential brain metabolism may be to produce no net change in $AVDO_2$.

Our pilot studies to assess the effect of propofol on $AVDO_2$ accord with those of others[1] and indicates that continuous $AVDO_2$ monitoring may provide a surrogate "end point" for the effects of metabolically active drugs upon the brain. The theoretical prospect of capitalising upon the enormous anti "excitotoxic" neuroprotective efficacy of glutamate antagonists, together with their putative sedative and analgesic properties, may make these agent the drugs of choice for severe head injury. Our dose escalation study aims to identify the optimal safe dose to achieve both these therapeutic aims, and $AVDO_2$ monitoring may be a useful technique to assess these aspects.

It is, however, imperative that attention be given to optimising technical factors with the Oximetric technique, and that the effects of both extracranial and intracranial haemodynamic factors be excluded prior to drawing conclusions about brain metabolism using this technique.

References

1. Andrews PJD, Dearden NM, Miller JD (1991) Comparison of thiopentone and propofol at two rates of administration in patients with severe head injury. Br J Anaesth 67: 212

2. Cruz J, Milne ME, Allan SJ, Alves WN, Gennarelli TA (1991) Continuous monitoring of cerebral oxygenation in acute brain injury: Assessment of cerebral haemodynamic reserves. Neurosurgery 129: 743–749

3. Muizelaar JP, Marmarou A, De Salles AF, Ward JD, Zinmerman RS, Zhong-Chao L, Choi SC, Young HF (1989) Cerebral blood flow and metabolism in severely head injured children. J Neurosurg 71: 63–71

4. Obrist WD, Langfitt TW, Jaggi JL, *et al* (1984) Cerebral blood flow and metabolism in comatose patients with acute head injury. Relationship to intracranial hypertension. J Neurosurg 61: 241–253

5. Rafferty C, Teasdale GM, Bullock R, Fitch W, Farling P (1993) Changes in jugular bulb oxygen saturation and associated changes in intracranial pressure. In: Avezaat C (ed) Proc. VIIIth International Conference on Intracranial Pressure. Springer, Berlin Heidelberg New York Tokyo (in press)

6. Robertson CS, Narayan RK, Gokaslan L, Pahwa R, Grossman R, Caram P, Allan E (1989) Cerebral arteriovenous oxygen difference as an estimate of cerebral blood flow in comatose patients. J Neurosurg 70: 222–230

Correspondence and Reprints: Dr. R. Bullock, Division of Neurosurgery, Medical College of Virginia, Virginia Commonwealth University, MCV Station, Box 631, Richmond, VA 23298-0631, U.S.A.

Consequences for Clinical Management

Acta Neurochir (1993) [Suppl] 59: 121–125
© Springer-Verlag 1993

Early and Late Systemic Hypotension as a Frequent and Fundamental Source of Cerebral Ischemia Following Severe Brain Injury in the Traumatic Coma Data Bank*

R. M. Chesnut[1], S. B. Marshall[1], J. Piek[3], B. A. Blunt[2], M. R. Klauber[2], and L. F. Marshall[1]

[1]Division of Neurological Surgery, [2]Department of Community and Family Medicine, University of California, San Diego Medical Center, San Diego, California, U.S.A., and [3]Neurochirurgische Klinik, Universität Düsseldorf, Düsseldorf, Federal Republic of Germany

Summary

The outcome from severe head injury (GCS $\leq$ 8 mmHg) was prospectively studied in patients from the Traumatic Coma Data Bank. We investigated the impact on outcome of hypotension (SBP < 90 mmHg) occurring from injury through resuscitation (early hypotension; N = 717) or in the Intensive Care Unit [ICU] (late hypotension; N = 493). Early hypotension occurred in 248 patients (34.6%) and was associated with a doubling of mortality (55% vs. 27%). If shock was present on admission, the mortality was 65%. These effects were independent of age, admission GCS motor score, presence of hypoxia, or associated severe extracranial trauma, suggesting that the influence of multiple system trauma in head injured patients is primarily due to associated hypotension. Late hypotension occurred in 156 of 493 patients (32%) and was the only hypotensive episode in 117 (24%). For 117 patients whose only hypotensive episode occurred in the ICU, 66% either died or were vegetative survivors, compared to 17% of patients who never suffered an hypotensive episode. Logistic regression modelling suggested that early and late shock were the most powerful independent predictors of mortality in this group of patients. These data demonstrate that hypotension is a common and devastating secondary brain insult in severe head injury patients, occurring not only during transport and resuscitation but also "right under our noses" in the ICU. We suggest that vigorous attention to eliminate or minimize such insults has the potential of markedly improving outcome from severe head injury.

Keywords: Severe head injury; outcome; systemic hypotension; coma data bank.

* This work was supported by the Traumatic Coma Data Bank (TCDB) under Contracts NO1-NS-3-2339, NO1-NS-3-2340, NO1-NS-3-2341, NO1-NS-3-2342, and NO1-NS-6-2305 from the National Institute of Neurological Disorders and Stroke (NINDS) and the "Wilhelm-Tönnis-Foundations" of the German Society for Neurosurgery. The TCDB Manual of Operations, which includes the TCDB data forms, is available from the National Technical Information Service (NTIS), U.S. Department of Commerce, 5285 Port Royal Road, Springfield, VA 22161 (NTIS Accession No. PB87 228060/AS).

Introduction

The role of ischemia in severe head injury has become widely accepted and a significant amount of research is now focused on halting or reversing the resultant underlying biochemical and metabolic abnormalities. The etiology of the ischemia, however, remains unclear. Much discussion centers about disrupted autoregulation, vasospasm, excitotoxicity, and inadequate cerebral perfusion pressure (CPP). The remarkably high incidence of ischemic damage in patients dying of their head injury as demonstrated by Graham et al.[2,3] suggests that there may well be a number of mechanisms interacting to produce ischemic injury in any given patient. It also appears, given the significant difficulties inherent in halting or reversing ischemic injury once it has been initiated, that a major focus of research should be the identification of early ischemic insults and the initiation of mechanisms to prevent their occurrence.

One potential and very fundamental source of ischemia is inadequate cerebral perfusion pressure due to systemic hypotension. Miller et al., as well as our group, have demonstrated that shock is common and highly detrimental during the early post-traumatic period[1,7,8]. A basic issue, therefore, becomes the role of systemic hypotension in producing ischemic insults to the injured brain during the entire period from injury through management in the neurosurgical intensive care unit. Systemic arterial pressure monitoring and volume resuscitation are basic treatment modalities that are present in some form in every intensive care unit in the world. Therefore, if systemic hypotension can be shown to be a predominant source of ischemic injury to the traumatized brain and a simple approach to preventing such insults can be devised, a pronounced improvement in the outcome from severe head injury could be forthcoming without the requirement of high technology devices or techniques. We have investigated the role of systemic hypotension in determining outcome from severe brain trauma from the time of injury through the patient's stay in the intensive care unit in order to shed further light on these issues.

Materials and Methods

The Traumatic Coma Data Bank (TCDB) is an NINDS collaborative project involving four clinical centers – the Medical College of Virginia at Richmond, the University of California at San Diego, the University of Virginia at Charlottesville, and the University of Texas Medical Branch at Galveston – with a coordinating center within the Biometry and Field Studies Branch, NINDS. These clinical centers prospectively studied all severely head injured patients admitted between April 1983 and April 1988. The operational definition of severe head injury was a Glasgow Coma Scale (GCS)[9] score of 8 or less occurring on admission (post-resuscitation) or during the ensuing 48 hours. In addition to data for the acute care course, prehospital information and rehabilitation follow-up results were collected. Patient outcome was determined by the last recorded Glasgow Outcome Scale (GOS)[5] score.

There were 1030 patients admitted to the TCDB. Of these, 284 were brain dead on admission, did not survive resuscitation, or suffered a gunshot wound to the brain, leaving 746 patients.

For our investigation of shock occurring from the time of injury through resuscitation, information on the prehospital course was insufficient to allow assessment in 29 patients, leaving 717. We defined "at hospital shock" as hypotension recorded from the time of arrival at the TCDB hospital through the end of resuscitation. In 18 patients, sufficient data was lacking on initial blood pressure or arterial blood gas results, leaving 699 patients for analysis of time of arrival data.

For our studies of "in-hospital shock", data analysis was complicated by significant overlap between resuscitation, ancillary diagnostic studies, emergent operative procedures, and the fact that the patient's arrival in the intensive care unit may have occurred anytime within a first eight hour shift. We therefore defined "late shock" as hypotension occurring in the intensive care unit setting anytime after the first eight hour ICU shift. In this group, 293 patients lacked one or more data values necessary for analysis, leaving 493 patients for final analysis.

For our studies, the operational definition of hypotension was a systolic blood pressure (SBP)≤90 mmHg. All patients had their injuries classified by system using the abbreviated injury scale (AIS). We define severe extracranial multiple trauma as any injury to the neck, thorax, diaphragm, abdomen, pelvis, spine, or extremities, with an AIS grade >3. Patients lacking such injuries were classified as lacking severe multiple trauma.

The independence of individual secondary brain insults, age, and severe multiple trauma in determining outcome trends was analyzed using the extended Mantel-Haenszel (M-H) chisquare

test[6]. Age was controlled in the M-H procedure using two groups: 0-39 years of age and 40 years of age and over. For summary purposes, the GOS score was collapsed from 5 to 3 categories in the tables; however, statistical testing using the GOS was performed using the full 5-point scale. Homogeneity of a relationship between categories of patients (controlling for hypoxia, hypotension, and severe multiple trauma) was tested using interaction terms in polychotomous logistic regression models[4]. The 5% level was used for statistical significance.

Results

Early Shock (Time of Injury through Resuscitation)

For this cohort, the median age was 25 years. The mean initial GCS score was 4. Sixty-one percent of patients were injured in motor vehicle accidents. Thirty-eight percent of patients were transfers to the TCDB hospital from another institution. The median transport time from injury to the receiving hospital was 0.9 hours.

Figure 1 presents the percentage of patients falling into each of three condensed GOS categories for cohorts experiencing no early shock, shock at anytime from injury through resuscitation, or presenting with shock upon admission. Of note is the marked increase in mortality from 27% for patients with no shock, to 50% for patients with shock at anytime, to 60% in patients presenting in shock. For the time period from injury through resuscitation, shock was experienced by 34.6% of patients. Analysis of the association of outcome with hypotension, controlling for hypoxia, age, and the presence or absence of associated severe multiple trauma, demonstrated hypotension to be extremely significant (p < 0.0001). If hypotension was combined with hypoxia (P_aO_2 < 60 mmHg), the combination was associated with significantly poorer outcome than that of hypotension alone (p = 0.021).

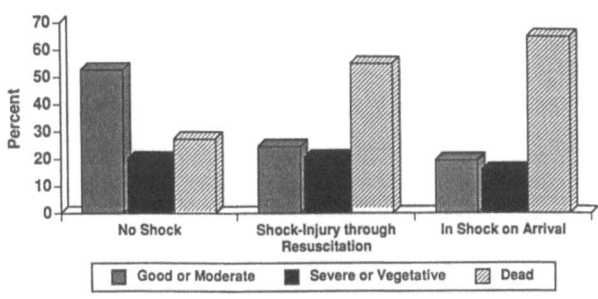

Fig. 1. Outcome as influenced by early shock (one or more episodes of systolic blood pressure ≤90 mmHg) for 717 patients in the Traumatic Coma Data Bank. Patients in shock on arrival are a subset of those experiencing shock during the time period from injury through resuscitation

Indeed, for patients presenting with hypoxia and hypotension at the time of admission, mortality was 75%.

Overall, there was a significant trend for poorer outcome in patients with associated severe multiple trauma (p = 0.0013). If hypoxia and shock, as secondary brain insults, are controlled for in the analysis in addition to age, however, severe multiple trauma is no longer statistically significant (p = 0.22). This suggests that the majority of the impact that severe multiple trauma has on outcome from severe head injury is attributable to the associated secondary brain insults, particularly hypotension.

Late Shock

Of the 493 patients available for this analysis, late shock occurred in a total of 32% of patients and was the only hypotensive insult in 24%. Figure 2 illustrates the increased morbidity and mortality associated with late shock as compared to patients experiencing no shock. Patients in the late shock category in this figure had no recorded hypotension through the first ICU shift. Note that, for patients surviving beyond the first ICU shift, 17% of patients experiencing no shock died or were vegetative, compared to 66% in patients suffering a late hypotensive episode. The difference in morbidity and mortality between these two groups was statistically significant (p < 0.001). These two groups were comparable with respect to age, sex, mechanism of injury, post-resuscitation GCS score (median = 6), and post-resuscitation GCS motor score. There was a significant difference between the two groups with respect to intracranial diagnosis. Logistic regression analysis analyzing this

interaction, however, found the presence or absence of late shock to remain a significant predictor of outcome when controlling for intracranial diagnosis (p < 0.001), which was also a significant predictor of outcome in this model (p < 0.001).

Logistic Regression Modelling of Excess Risk

Logistic regression analysis of these data suggested that the presence of early shock was associated with a 15-fold excess mortality, late shock was associated with an 11-fold excess mortality, and intracranial diagnosis was associated with a 5.2-fold excess mortality. Age was associated with a 4% increase in predicted mortality per year of age. In this regression analysis, multiple trauma was not a statistically significant independent predictor of increased mortality (p = 0.86).

Discussion

Between 1984 and 1988, over 40% of severely head injured patients admitted to the TCDB via four academic medical centers with specific clinical and basic science interest in brain trauma suffered one or more hypotensive insults associated with a profound increase in their morbidity and mortality. These episodes of shock, although probably associated with other secondary insults, severe extracranial multiple trauma, and intracranial diagnosis, are statistically independent of these other factors with respect to their influence on outcome. Given the fundamental nature of the insult and the ready availability of mechanisms for monitoring and treating it in any health care setting, a potential mechanism for dramatically improving the outcome from severe head injury exists on a widespread basis.

There are several significant shortcomings with this investigation, generally related to the broad scope of the study design. First of all, we do not know the duration of any of these ischemic insults, and this information is not available in our data set. With respect to the early shock groups, we do know that a significant number of the hypotensive insults were relatively short, on the order of 15 minutes or less.

Because of the method of data collection, we are also unable to define the medical circumstances surrounding the hypotensive episode with sufficient detail for meaningful analysis. We are, therefore, unable to comment on the number of episodes that

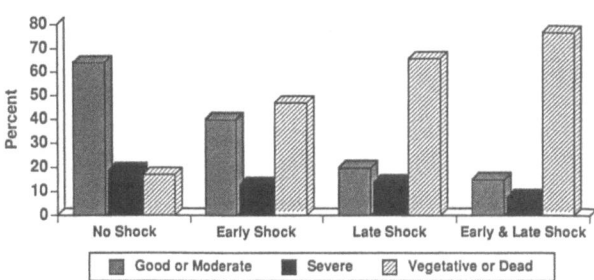

Fig. 2. Outcome as influenced by in-hospital shock (one or more episodes of systolic blood pressure ≤90 mmHg) for 493 patients in the Traumatic Coma Data Bank. Early shock is hypotension on arrival at the TCDB hospital. Late shock encompasses the patient's stay in the ICU beginning after the first shift

were directly resultant from confounding medical conditions. Given the time course of the shock, however, with the median time of occurrence of late shock being 31 hours, in combination with our own experience with a significant proportion of the patients included in this cohort, it is known that an unquantified majority of late hypotensive episodes cannot be attributed to complicating medical conditions such as sepsis.

A final serious shortcoming of this preliminary report is the absence of cerebral perfusion pressure data for the episodes of late shock. This data actually is available in the data set, but, due to the method in which the data was collected and encoded, extracting this information with a high degree of precision is arduous and has not yet been completed. Certainly, a SBP of 90 mmHg suggests a mean arterial pressure in the range of 65–70 mmHg. In a patient with borderline intracranial hypertension at 20 mmHg, this would give a cerebral perfusion pressure on the order of 45–50 mmHg. These estimates are in accordance with the early stages of our data analysis which suggests that the mean CPP in these patients will be in the upper 30s or lower 40s.

Over the past several years, significant concern has arisen as to the proper minimum acceptable CPP in the head trauma patient. Although 50 mmHg is generally felt to be the lower end of the physiologic range in the normal patient, there is a general feeling (which is becoming supported by a burgeoning mass of clinical data) that the cutoff point should be elevated (to the range of 70 mmHg or more) after brain injury. In this light, a SBP of 90 mmHg would generally represent a CPP that is unacceptable even in the absence of intracranial hypertension.

Intercurrent with our TCDB findings regarding the frequency and impact on outcome of late shock, we have reviewed the courses of a more recent group of patients treated in our ICU. In the recent past, the assignation of 70 mmHg as the minimum acceptable CPP has increasingly become the standard of care at UCSD Medical Center. One notable finding in patients managed in keeping with this elevated CPP level is that the incidence of hypotension (by our definition of a SBP of ≤90 mmHg) has decreased by approximately 60%. In addition, when these hypotensive episodes are not related to significant associated medical disturbances, they are almost uniformly very short in duration. Unfortunately, hypotensive episodes associated with medical disorders such as sepsis remain rather refractory to management and tend to be more protracted and severe.

It is notable, albeit not surprising, that the elevation of the minimum acceptable CPP to a higher level appears to be associated with a significant decrease in the incidence of systemic hypotension. Given the data presented above, this is quite likely a significant factor in the efficacy of such CPP management in improving outcome from severe head injury. This factor certainly needs to be considered and, optimally, compared to a control group, when any beneficial effects of elevating the minimum acceptable CPP are attributed to other mechanisms such as improvements in ICP control.

In summary, if systemic hypotension is a significant factor in the poor outcome from severe head injury, it is an issue that can be easily and straightforwardly addressed at any level or period in the care of the severely head injured patient. Increased awareness of the remarkable susceptibility of the traumatized brain to any period of systemic hypotension, even when of short duration, coupled with preventive measures such as vigorous volume resuscitation in head injured patients and management protocols that call for efficacious and instantaneous correction of any episode of systemic hypotension regardless of the etiology, have the potential of significantly improving the outcome from severe head injury. Finally, it is suggested that the elevation of the minimum acceptable CPP to 70 mmHg might serve as significant prophylaxis against episodes of systemic hypotension. This must be taken into the analysis when assessing the effect of such therapy on outcome statistics. The elevation of the minimum acceptable CPP as part of the treatment of the severely head injured patient should be considered on the basis of its prophylaxis against hypotension alone.

References

1. Chesnut RM, Marshall LF, Klauber MR, et al (1992) The role of secondary brain injury in determining outcome from severe head injury. J Trauma 34: 216–222
2. Graham DI, Adams JH, Doyle D (1978) Ischaemic brain damage in fatal non-missile head injuries. J Neurol Sci 39: 213–234
3. Graham DI, Ford I, Adams JH, et al (1989) Ischaemic brain damage is still common in fatal non-missile head injury. J Neurol Neurosurg Psychiatry 52: 346–50

4. Hosmer DW, Lemeshow S (1989) Applied logistic regression. J. Wiley, New York
5. Jennett B, Bond M (1975) Assessment of outcome after severe brain damage: A practical scale. Lancet 1: 480–484
6. Mantel N (1963) Chi-square tests with one degree of freedom; extensions of the Mantel-Haenszel procedure. J Am Stat Assoc 58: 690–700
7. Miller JD, Becker DP (1982) Secondary insults to the injured brain. J Royal Coll Surg (Edinburgh) 27: 292–298
8. Miller JD, Sweet RC, Narayan R, *et al* (1978) Early insults to the injured brain. J Am Med Assoc 240: 439–442
9. Teasdale G, Jennett B (1974) Assessment of coma and impaired consciousness: A practical scale. Lancet 2: 81–84

Correspondence and Reprints: R. M. Chesnut, M.D., Division of Neurological Surgery, University of California, San Diego Medical Center, San Diego, CA 92107-8893, U.S.A.

Alexander Baethmann, Oliver Kempski, Ludwig Schürer (eds.)

Mechanisms of Secondary Brain Damage
Current State

1993. 76 figures. VIII, 165 pages.
Cloth DM 160,–, öS 1120,–
Reduced price for subscribers to "Acta Neurochirurgica":
Cloth DM 144,–, öS 1008,–
ISBN 3-211-82421-9

(Acta Neurochirurgica / Supplementum 57)

This volume presents an interdisciplinary discussion on current knowledge of mechanisms of secondary brain damage evolving from cerebral trauma or ischemia. Contributions reach from the basic sciences to the daily clinical practice. Specific topics focus on functional analysis of the brain by NMR-spectroscopy or PET-scanning, as well as on quantitative histomorphological assessment of primary vs. secondary lesions in neurotraumatology. Other chapters are concerned with the acid-base regulation in cerebral ischemia, the current understanding of clinically relevant mediator compounds enhancing tissue damage in trauma or ischemia, in particular focusing on the amino acid glutamate. Valuable information is dealing with recent investigations on the recovery potential of the brain from ischemia in men, or the most modern treatment modalities of cerebral resuscitation after cardiac arrest, combining among others cardiopulmonary bypass, hemodilution, and hypothermia. Further highlights are discussions of the present state of prehospital emergency care, including logistics, training of personnel, and availability of specific trauma centers and, last but not least, evaluations of the clinical management of intracranial hypertension and brain tissue acidosis in severe head injury.

Springer-Verlag Wien NewYork

Sachsenplatz 4–6, P.O.Box 89, A-1201 Wien · 175 Fifth Avenue, New York, NY 10010, USA
Heidelberger Platz 3, D-14197 Berlin · 37-3, Hongo 3-chome, Bunkyo-ku, Tokyo 113, Japan

Hans.-J. Reulen, Alexander Baethmann, Joseph Fenstermacher,
Anthony Marmarou, Maria Spatz (eds.)

Brain Edema VIII
Proceedings of the Eighth International Symposium Bern, June 17-20, 1990

1990. 203 figures. XIII, 416 pages.
Cloth DM 250,–, öS 1750,–
Reduced price for subscribers to "Acta Neurochirurgica":
Cloth DM 225,–, öS 1575,–
ISBN 3-211-82240-2

(Acta Neurochirurgica / Supplementum 51)

The book is a compilation of papers presented at the Eighth International Symposium on Brain Edema, held from June 17-20, 1990, in Bern, Switzerland. The Symposium explored clinical as well as basic science aspects of this topic. Clinicians and scientists from many fields of neurosciences contributed to the state-of-the art presentations and discussions.

The papers in this volume are grouped for the reader´s convenience. Beside chapters on the pathophysiology the papers are focused around the major disease processes associated with brain edema: tumors, trauma, ischemia, hydrocephalus, hypertension, infection, pseudotumor cerebri etc. This allowed to group together the neuropathology, pathophysiology including cellular and molecular phenomena, clinical findings, in vivo diagnosis with MRI as well as therapeutic aspects of a given entity. A major issue was the role of biochemically active mediator substances in opening the blood brain barrier and development of edema. Methods of their specific inhibition may become the most important and effective therapeutic interventions in the future.

Editors, authors and Springer-Verlag have made a great effort to publish this volume - the eighth in the series - within 6 month after the Symposium. Thus, the volume reflects an up-to-date knowledge of the disease and the possible modes of treatment as well as new advances in innovative research. The book provides important information for all those who are clinically or scientifically involved with this disease.

Sachsenplatz 4–6, P.O.Box 89, A-1201 Wien · 175 Fifth Avenue, New York, NY 10010, USA
Heidelberger Platz 3, D-14197 Berlin · 37-3, Hongo 3-chome, Bunkyo-ku, Tokyo 113, Japan

Hans-Jakob Steiger

Pathophysiology of Development and Rupture of Cerebral Aneurysms

1990. 51 partly coloured figs. IX, 57 pages.
Cloth DM 77,–, öS 540,–
Reduced price for subscribers to "Acta Neurochirurgica":
Cloth DM 70,–, öS 486,–
ISBN 3-211-82192-9

(Acta Neurochirurgica / Supplementum 48)

Based on a series of experimental studies, this volume provides a new perspective of the pathophysiological factors responsible for initial development, growth and eventual rupture of cerebral saccular aneurysms. The blood flow patterns in cerebral arteries are analysed in the first part and the stresses responsible for the initial formulation of aneurysms are derived. The circulation inside already formed aneurysms is subsequently dealt with and a concept of the anatomical relations of saccular aneurysms is proposed, which is based on the intraaneurysmal blood flow. The properties of aneurysmal tissue are then analysed and an understanding of the mechanisms involved in growth and rupture on aneurysms is developed in the last part. The text addresses clinical neurosurgeons as well as basic scientists. The concepts developed have potential impact on the management of patients harbouring ruptured or unruptured aneurysms.

Springer-Verlag Wien New York

Sachsenplatz 4–6, P.O.Box 89, A-1201 Wien · 175 Fifth Avenue, New York, NY 10010, USA
Heidelberger Platz 3, D-14197 Berlin · 37-3, Hongo 3-chome, Bunkyo-ku, Tokyo 113, Japan